An Introduction to Quantitative Finance

STEPHEN BLYTH

Professor of the Practice of Statistics, Harvard University
Managing Director, Harvard Management Company

OXFORD
UNIVERSITY PRESS

Great Clarendon Street, Oxford, OX2 6DP,
United Kingdom

Oxford University Press is a department of the University of Oxford.
It furthers the University's objective of excellence in research, scholarship,
and education by publishing worldwide. Oxford is a registered trade mark of
Oxford University Press in the UK and in certain other countries

First Edition published 2014

Published in the United States of America by Oxford University Press
198 Madison Avenue, New York, NY 10016, United States of America

British Library Cataloguing in Publication Data
Data available

Library of Congress Control Number: 2013941968

ISBN 978–0–19–966659–1

Printed and bound by CPI Group (UK) Ltd, Croydon, CR0 4YY

For Athena, Sophia and Anita

PREFACE

This book is based on *Applied Quantitative Finance on Wall Street*, an upper level undergraduate statistics course that I have taught for several years at Harvard University. The students taking the class are typically undergraduate concentrators in mathematics, applied mathematics, statistics, physics, economics and computer science; master's students in statistics; or PhD students in quantitative disciplines. No prior exposure to finance or financial terminology is assumed.

Many students who are considering a career in finance study this material to gain an insight into the machinery of financial engineering. However, some of my students have no interest in a financial career, but simply enjoy probability and are motivated to explore one of its compelling applications and investigate a new way of thinking about uncertainty. Others may be curious about quantitative finance due to the regulatory and policy-making attention the industry has received since the financial crisis.

The book concerns financial derivatives, a derivative being a contract or trade (or bet, depending on your prejudices) between two entities or counterparties whose value is a function of—derives from—the price of an underlying financial asset. We define various derivative contracts and describe the quantitative and probabilistic tools that were developed to address issues encountered by practitioners as markets developed. The book is steeped in practice, as the methods we explore only developed to such an extent because the markets themselves grew exponentially fast. Whilst we develop theory here, this book is not theoretical in the 'existing only in theory' sense. The products we consider are traded in significant size and have meaningful economic impact.

Probability provides the key tools for analysing and valuing derivatives. The price of a stock or a bond at some future time is a random variable. The payout of a derivative contract is a function of a random variable, and thus itself another random variable. We show that the price one should pay for a derivative is closely linked to the expected value of its payout, and that suitably scaled derivative prices are martingales, which are fundamentally important objects in probability theory. The book focuses largely on interest rate derivatives—where the underlying financial variable is an interest rate—for three reasons. First, they constitute by far the largest and most economically important derivative market in the world. Secondly, they are typically the most challenging mathematically. Thirdly, they have generally been less well addressed by finance textbooks.

Background and motivation

The structure and content of the book are shaped by the experiences of my career in both academia and on Wall Street. I read mathematics as an undergraduate at Cambridge University where I was fortunate to be taught probability by both Frank Kelly and David

Williams. I obtained a PhD in Statistics at Harvard University with Kjell Doksum before teaching for two years at Imperial College London, where I lectured the introductory probability course for mathematics undergraduates.

My move to Wall Street was catalysed indirectly by President Clinton's decision to cancel the superconducting super collider in 1993. Two Harvard housemates, both theoretical physicists, saw the academic job market for physicists collapse overnight and both quickly found employment at Goldman Sachs in New York. Their assertion that derivative markets (whatever in fact they were) were 'pretty interesting' and mathematically challenging convinced me to contemplate a move from my nascent academic career.

I started at HSBC in London, hired directly by a Cambridge PhD mathematician. On my first day at work I was tasked to calculate the expected value of a function of bivariate normal random variables. I moved subsequently to Morgan Stanley in New York where I became managing director and the market maker for US interest rate options. Throughout my stint at Morgan Stanley I worked with another Cambridge mathematician. Together we developed a training program for new analysts, motivated by mathematical and probabilistic challenges and subtleties we encountered in the rapidly developing interest rate options markets of the late 1990s. Several of the problems from that mini-course have found their way into this book. Indeed much of the theory I develop is motivated by a desire to provide a suitably rigorous yet accessible foundation to tackle problems I encountered whilst trading derivatives. This fundamental motivation led me to create the course at Harvard and subsequently develop this book, which I believe combines an unusual blend of derivatives trading experience and rigorous academic background.

After a spell in London running an interest rate proprietary trading group at Deutsche Bank, I rejoined Harvard in 2006 where I currently wear two hats: head of public markets at the Harvard Management Company, the subsidiary of the university responsible for managing the endowment; and Professor of the Practice of statistics at Harvard University, where I teach the course. My role at the endowment includes responsibility for Harvard's investment activities across public bond, equity and commodity markets. This ongoing presence at the heart of financial markets—I continue to trade each day derivative contracts explained in the text—means that the course, and thus I hope the book, retains immediacy to its field. In particular, the financial crisis of 2008–2009 challenged many basic foundational assumptions both academics and practitioners had taken for granted and I detail several of these in the exposition, some surprisingly early.

The Harvard statistician Arthur Dempster made the distinction between 'procedural' statistics, involving the rote application of procedures and methods, and 'logicist' statistics in which reasoning about the problem at hand—using subject-matter knowledge and judgment in addition to quantitative expertise—plays a central role. This book is written in the logicist tradition, interweaving experience and judgment from the real world of finance with the development of appropriate quantitative techniques.

It is often not obvious to an outsider which issues are particularly important to traders and investors in the Wall Street derivative markets. Whilst everything in this book is to some extent motivated by the practical world, we highlight areas where the degree of market relevance is not immediately apparent. We also include market anecdotes and tales from the Street to provide insights and richness into the material.

This book has been motivated by and distilled from my experience on Wall Street, and also by my love for probability. In addition to wanting to understand elements of quantitative finance in practice, students may wish to read the book simply because they have enjoyed their introduction to probability and have a desire to see how it can be applied. Whilst there are many ways to approach financial mathematics, I find the probabilistic route both appealing and fulfilling. There are highly intuitive elements to the theory, for example prices being expected values. Furthermore, we introduce in a straightforward manner powerful concepts such as numeraires and martingales that drive much of modern theory, and discover that there are subtleties in various unexpected places that demand precise and non-trivial understanding.

Prerequisites

The sole prerequisite for mastering the material in this book is a solid introductory undergraduate course in probability (represented by Statistics 110 at Harvard, Part IA Probability at Cambridge University, or Probability & Statistics I at Imperial College). Familiarity is required with: discrete and continuous random variables and distributions; expectation, in particular expectation of a function of a random variable; conditional expectation; the binomial and normal distributions; and an elementary version of the central limit theorem. The single-variable calculus typically associated with such a course—integration by parts, chain rule for differentiation and elementary Taylor series—is used at several points. Exercises at the end of this preface give a sense of the probability prerequisite.

The book is otherwise self-contained and in particular requires no additional preparation or exposure to finance. The necessary financial terminology is introduced and explained as required. I tell my students that the edifice of quantitative finance was largely built by pioneers from academic physics or mathematics who entered Wall Street with no knowledge of finance and having read no finance books, and who still have not.

This is not a book on economics. There are many excellent economics texts on capital markets, corporate finance, portfolio theory and related matters, which complement this material. However, derivative markets are a world largely populated by traders and mathematicians, not economists. Nor is this a book on time series methods or prediction. As we see early on, derivative pricing—even that concerning the price of a forward contract which is an exchange of an asset in the future—is largely unrelated to prediction.

The mathematical prerequisite is modest and no more extensive than that for an introductory undergraduate probability course, although adeptness at logical quantitative reasoning is important. Whilst we build an appropriately rigorous base for our results, this is not a theoretical mathematics book. By developing most of our theory using discrete-time methods, we can address important issues in finance without exposure to stochastic processes, Ito calculus, partial differential equations, Monte Carlo or other numerical methods, all of which play important roles in continuous-time financial mathematics. We raise signposts to areas of further study in the rough guide to continuous time that concludes the book.

I have necessarily had to be ruthless with material I have excluded in order to keep the book appropriate to a one-semester course of approximately 33 lecture hours. The book is selective rather than encyclopaedic and moves at a lively pace. This enables students to be exposed to powerful theory and substantive problems in one semester. Throughout the book I raise signposts to more advanced topics and to different approaches to the material which naturally arise from the exposition. Even an encyclopaedic tome cannot cover all dimensions of derivative markets and associated theory.

Outline

The book is organized in five parts. The first, short part contains some preliminaries regarding interest rates and zero coupon bonds, and a very brief sketch of assets.

The bulk of the material is contained in Parts II-IV. In Part II, we define and explore basic derivative products including forwards, swaps and options. Our first derivative—the forward contract—is introduced and we consider methods for valuation and pricing which we discover do not depend on any probabilistic modelling. Recent violations of traditional assumptions during the financial crisis are soon encountered. We introduce elementary interest rate derivatives, such as forward rate agreements (FRAs) and swaps, and define futures, the first contract whose price depends on distributional assumptions of the underlying asset. We state formally the no-arbitrage principle, a foundation of quantitative finance, which makes precise the arguments to which we have already appealed. Recent empirical challenges to the principle of no-arbitrage manifested during the financial crisis are discussed. We then introduce and define call and put options, key building blocks of modern derivative markets. We investigate option properties and establish model-independent bounds on their price. We also review options on forwards, an area that can still cause puzzlement on Wall Street.

Part III focuses on the key pricing arguments of replication and risk-neutrality. We formalize arguments regarding replication and hedging on a binomial tree, and define risk-neutral valuation in this setting. We show that prices are appropriately scaled expected values under the risk-neutral probability distribution, a result which leads to the fundamental theorem of asset pricing for the binomial tree, the most powerful result of the book. Introducing in a natural manner the concepts of numeraire and martingale, we give a general form of the fundamental theorem, which states that no-arbitrage is equivalent to the ratios of prices to a numeraire being martingales. Bridging in a simple manner to continuous time, we take limits using the central limit theorem and move to a continuous case where expectation under a lognormal density immediately gives the Black–Scholes formula. We review its properties and introduce the concepts of delta and vega. The key duality between option prices and probability distributions is then explored in several ways.

Part IV develops the understanding of interest rate options, the largest, most important and most mathematically challenging of options markets. No book covers the material particularly well and the fabric of these markets is typically only seen clearly by its practitioners. We introduce and value interest rate caps, floors and swaptions, using an elegant choice of numeraire. We then define Bermudan swaptions and derive arbitrage bounds for their value, and construct trades using cancellable swaps. Two further topics in interest rate options are discussed: libor-in-arrears and general convexity corrections, in particular the details of one of the great trades in derivatives history; and a brief description of the BGM model and its volatility surfaces, which underlies much of current interest rate modelling.

In the concluding Part V, we provide a brief introductory sketch of the continuous-time analogue of the theory developed in the book, and raise signposts to the partial differential equation approach to mathematical finance.

The book includes substantive homework problems in each chapter (other than Part V) which illustrate and build on the material. The aim is to make the book fulfilling, challenging and entertaining, and to convey the immediacy and practical context of the material we cover.

EXERCISES

The following exercises provide an indication of the level of probability proficiency that is required to master the material in this book.

1. **The normal distribution and expectation**
   Let $X \sim N(\mu, \psi^2)$ and K be a constant. Calculate
   (a) $E(I\{X > K\})$ where I is the indicator function.
   (b) $E(\max\{K - X, 0\})$.

   Hint Use $(K - X) = (K - \mu) - (X - \mu)$.

   (c) $E(e^{tX})$.
   Leave answers where appropriate in terms of the normal cumulative distribution function $\Phi(\cdot)$.

   Note In the language of derivatives (a) is essentially the price of a ***digital call option*** and (b) is the price of a ***put option***. We encounter these in Chapter 7.

2. **The lognormal distribution**
   A random variable Y is said to be lognormally distributed with parameters μ and σ if $\log Y \sim N(\mu, \sigma^2)$. We sometimes write $Y \sim \text{lognormal}(\mu, \sigma^2)$.
   (a) Using your answer to Question 1(c), calculate $E(Y)$ and $\text{Var}(Y)$.
   (b) Suppose $\mu = \log f - \frac{1}{2}\sigma^2$ for a constant $f > 0$. Calculate $E(Y)$ and show that $\text{Var}(Y) = f^2\sigma^2(1 + \frac{\sigma^2}{2} + \frac{\sigma^4}{6} + \text{higher order terms})$.

3. **Conditional expectation**
A sequence of random variables $X_0, \dots, X_n, \dots$ is defined by $X_0 = 1$ and $X_n = X_{n-1}\xi_{n-1}$, where ξ_i, for $i = 0, 1, \dots$, are independent and identically distributed with

$$\xi_i = \begin{cases} 1 + u & \text{with probability } p \\ 1 + d & \text{with probability } 1 - p, \end{cases}$$

where $u > d$.

(a) Calculate $E(X_n \mid X_{n-1})$.

(b) Let $Y_n = \frac{X_n}{(1+r)^n}$ for a constant $r > 0$. Find in terms of p, u and d the value of r such that $E(Y_n \mid Y_{n-1}) = Y_{n-1}$. Hence find in terms of u and d the range of such possible values for r.

(c) Suppose $E(Y_n \mid Y_{n-1}) = Y_{n-1}$ for all n. Use any result concerning conditional expectation to prove that $E(Y_n \mid Y_m) = Y_m$ for all $n > m \geq 0$.

Note The result in (c) is essentially the definition that Y_n is a ***martingale***, which we encounter in Chapter 9.

4. **The normal cumulative distribution function**
Show that the first three non-zero terms of the Taylor series for the normal cumulative distribution function $\Phi(x)$ around zero are

$$\Phi(x) = \frac{1}{2} + \frac{x}{\sqrt{2\pi}} - \frac{x^3}{6\sqrt{2\pi}}.$$

ACKNOWLEDGMENTS

I am grateful to the many colleagues, fellow practitioners and students who read earlier versions of this book, including Raj Bhuptani, Joseph Blitzstein, John Campbell, Howard Corb, Kjell Doksum, Mohamed El-Erian, Jessy Hwang, Amol Jain, Bo Jiang, Frank Kelly, David Land, David Moore and Andy Morton, plus the entire Tufnell Park team. They contributed many helpful and insightful comments.

I was fortunate to have been taught probability by Frank Kelly and David Williams at Cambridge University, both of whom inspired in me a love for the subject that has never faded, and statistical theory and reasoning by Kjell Doksum and Arthur Dempster at Harvard University. These four professors provided me the foundational underpinnings from which I was able to tackle the world of quantitative finance. I learnt finance largely on the job in London and New York, in particular from two fine Cambridge mathematicians, Dan George and David Moore. I sat next to David for six years and gained from him an appreciation of the subtleties of financial markets and the theory required to understand them.

I am particularly grateful to Jane Mendillo who kindly allowed me to spend time in the classroom amid our responsibilities at the endowment, and who has been an unwavering supporter of all my endeavours at the University, to Xiao-Li Meng for welcoming me so warmly to the faculty of the Statistics department, and to Mohamed El-Erian for bringing me back to Harvard in the first place.

Keith Mansfield at OUP has been a strong advocate for this project from the start, and has guided me wisely through its evolution. Willis Ho, Kelly Sampson and Alex Zhu expertly helped produce the figures in the book, and Johnny Roche provided technical support throughout.

Finally, I thank all the students who have taken my class at Harvard, out of which this book grew. They have made teaching an unqualified joy.

CONTENTS

PART I
Preliminaries

1

Preliminaries

One of the interview questions I was asked when applying for my first job in finance was, 'Would you rather have one dollar today or one dollar in one year's time?' The question was asked abruptly without any polite preamble, as if to throw me off guard. However, the answer is clearly that one would rather have the dollar today since one can, amongst other things, deposit the dollar at a bank and receive interest for the year. This is preferable to receiving a dollar in a year's time, provided interest rates are not negative. The bulk of this preliminary chapter is concerned with the mechanics of interest rates, compounding and computing the value today of receiving a dollar or other unit of currency at some date in the future.

Note Almost universally throughout quantitative finance, interest rates are assumed to be non-negative. However, finance in practice has the habit of throwing up unexpected complexity even in the simplest of settings, and interest rates are no exception. Figure 1.1 shows the graph of the two-year Swiss interest rate (precisely, the two-year government bond yield, which we encounter in Chapter 4). We observe that this rate has been negative for non-negligible periods. How can this be so? Surely we can simply keep hold of our Swiss francs, and not deposit them at negative rates, thus establishing an effective floor of 0% on interest rates? In practice, it is not easy to keep billions of Swiss francs in safes (or under the mattress). The Swiss national bank—as part of its aim to reduce the strength of its currency during a particularly intense episode of euro concerns in late 2010—penalized all deposits of Swiss francs by way of negative interest rates. However, throughout this book we assume, unless stated otherwise, that all interest rates are non-negative, keeping in mind that markets have a tendency to challenge even fundamentally sound assumptions.

1.1 Interest rates and compounding

Suppose we deposit (or invest) an amount $\$N$ at a rate r per annum, compounded annually. Then we have after one year $\$N(1 + r)$, and after T years an amount $\$N(1 + r)^T$. Throughout the book we will express interest rates as decimals, so r is a rate of $100r\%$.

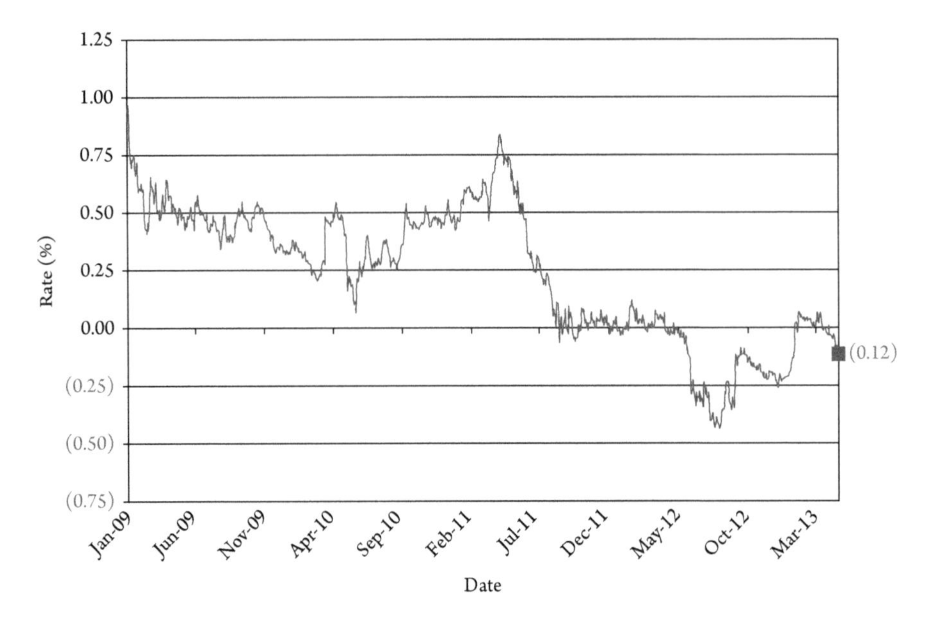

Figure 1.1 Two-year Swiss government bond yield

N is called the ***notional*** or ***principal***. For simplicity, we will usually assume that $N = 1$ unless otherwise stated, as we can always multiply through by N. We will usually also omit the $ or other currency symbol.

Note Concern amongst regulators often refers to the large size of outstanding notional, even though we will see later that payments on most derivative contracts relate to rN, or indeed $(r_1 - r_2)N$, where r_1 and r_2 are two different interest rates. These payments are thus usually of smaller order of magnitude than the notional.

If we invest 1 at r compounded semi-annually, we have $(1 + r/2)$ after six months and $(1 + r/2)^{2T}$ after T years.

If we invest 1 at r compounded m times per annum, we have at time T an amount $(1 + r/m)^{mT} \to e^{rT}$ as $m \to \infty$. Therefore, a unit amount compounded continuously at r becomes e^{rT} after T years.

Result *Suppose the continuously compounded rate for period T is r. Then the equivalent rate r_m with compounding frequency m is*

$$r_m = m(e^{(r/m)} - 1).$$

Proof The result follows immediately by noting that

$$e^{rT} = (1 + r_m/m)^{mT}. \qquad \blacksquare$$

For example, 100 invested at 5% with annual compounding for two years becomes 110.25 at $T = 2$, equivalent to a 4.879% rate with continuous compounding. 100 invested at 5% with continuous compounding for two years grows to 110.52.

The ***money market account***, M_t, is the value at time t of 1 invested at interest rate r. Thus $M_0 = 1$, and if r is the constant continuously compounded rate, $M_t = e^{rt}$. It is possible to generalize to when r varies with time or is random. The money market account plays an important role when we discuss the fundamental theorem of asset pricing in Chapter 9.

1.2 Zero coupon bonds and discounting

A ***zero coupon bond*** (ZCB) with maturity T is an asset that pays 1 at time T (and nothing else). The ZCB is such a fundamental object that we introduce it before we define bond or coupon!

Let $Z(t, T)$ be the price at current time t of a ZCB with maturity T, for $t \leq T$. By definition $Z(T, T) = 1$.

Result *Suppose the continuously compounded interest rate from t until T is a constant r. Then we have*

$$Z(t, T) = e^{-r(T-t)}.$$

Proof At time t, consider the following two portfolios.

Portfolio **A**. One zero coupon bond with maturity T.

Portfolio **B**. $e^{-r(T-t)}$ of cash, which we deposit at r.

Both portfolios are worth 1 at time T, and therefore they must be worth the same at time t. ■

Similarly, if r_A is the annually compounded rate, then

$$Z(t, T) = \frac{1}{(1 + r_A)^{(T-t)}}.$$

Note The logic of this argument, namely that if two portfolios are worth the same at some future time T then they are worth the same at current time $t \leq T$, may seem obvious—or is it?—and we adopt similar arguments in Chapter 2. We formalize the logic behind the argument in Chapter 6 using the axiom of no-arbitrage, and see recent empirical challenges to this logic presented by the financial crisis.

The values $Z(t, T), 0 \leq t \leq T$, are also known as ***discount factors***, or present values. A known cashflow of K at time T is worth $KZ(t, T)$ at time t. This is often termed 'discounting cashflows back to today' or 'present-valuing'.

Note Notation in quantitative finance often varies across authors and the ZCB price is a good example, which is variously denoted in textbooks by B for 'bond', P for 'present value' or D for 'discount factor'. In this book we always use Z for 'zero coupon'. We use W for the standard normal random variable to avoid confusion with the use of Z.

Given the values $Z(t,T)$, we can recover interest rates. For example, the unique interest rate r_A such that

$$Z(t,T)(1+r_A)^{(T-t)} = 1$$

is known as the annual zero coupon rate or ***zero rate*** for period t to T. Similarly the rate r such that

$$Z(t,T)e^{r(T-t)} = 1$$

is the ***continuous zero rate*** for period t to T.

Note that r is typically not constant across time. For example, we should rigorously write an expression like

$$Z(t,T)(1+r_{A,t,T})^{(T-t)} = 1$$

where we specify the compounding frequency and the term of the interest rate. However, usually the compounding frequency will be suppressed and deduced from context, and often the term will be as well. When we introduce libor and swap rates in Chapters 3 and 4, we establish a more precise notation for interest rates.

It is important to be comfortable moving between discrete and continuous compounding. The market is always effectively discrete since daily compounding is the highest practical frequency, but continuous compounding (exponential growth) is often mathematically more convenient.

Note that interest rates are always expressed as per-annum rates, regardless of compounding frequency or term.

1.3 Annuities

An ***annuity*** is a series of fixed cashflows C at specified times T_i, for $i = 1, \ldots, n$.

The value V at current time $t \leq T_1$ of an annuity is given by

$$V = C\sum_{i=1}^{n} Z(t,T_i).$$

Result *Suppose an annuity pays* 1 *every year for M years, and that annually compounded zero rates are r for all maturities* $T = 1, \ldots, M$. *Then*

$$V = \sum_{i=1}^{M} \frac{1}{(1+r)^i} = \frac{1}{r}\left(1 - \frac{1}{(1+r)^M}\right).$$

Proof We calculate the sum of this simple geometric progression by noting that

$$\frac{1}{(1+r)}V - V = \frac{1}{(1+r)^{(M+1)}} - \frac{1}{(1+r)}. \qquad \blacksquare$$

Geometric progressions of this form crop up throughout finance. Note that if $r = 0$ then $V = M$, and as $r \to \infty$, $V \to 0$. In a typical example, say with interest rates of 4% and $M = 10$, then $V = 8.11$.

For example, in the US a $100 million 'Powerball' lottery jackpot is typically structured as an annuity paying, say, $4 million per year for 25 years. With interest rates at 3%, this jackpot is actually only worth $69.7 million.

We revisit annuities when we consider the fixed leg of interest rate swaps in Chapter 4.

1.4 Daycount conventions

We saw that when compounding m times per annum, r/m of notional was paid in interest each $1/m$ years. The fraction $1/m$ is called the ***accrual factor***, often denoted by α. Suppose we have quarterly payments, so $m = 4$. In practice, α is not exactly 0.25 since a year does not divide easily into four equal parts. For example, Table 1.1 shows the number of days between these quarterly dates in 2011–2012.

Due to the idiosyncratic nature of how markets developed, there are many different market conventions for calculating the actual accrual fraction used when determining interest payments for a particular period. For example, the 'actual three-sixty' (act/360) daycount convention counts the actual number of days in the period and divides by 360, not 365. This was originally designed to help calculation. The 'thirty three-sixty' (30/360) convention assumes that there are 30 days in each month and was designed to produce equal amounts of interest payment each month or quarter. Table 1.2 gives a quick summary of various conventions and an example of the accrual factor that each gives rise to.

The interest rate quoted with an act/360 daycount will be lower than the equivalent rate for the same period quoted with an act/365 or 30/360 daycount. For example, if on 16

Table 1.1 Number of days between quarterly dates.

Period	**Days**
16 March 2011 to 16 June 2011	92
16 June 2011 to 16 September 2011	92
16 September 2011 to 16 December 2011	91
16 December 2011 to 16 March 2012	91

Table 1.2 Daycount conventions and accrual factors.

Daycount convention	α for 16 December 2011–16 March 2012
act/365	91/365
act/act	15/365 + 76/366
act/360	91/360
30/360	1/4

December 2011 the act/365 rate for three months is 5%, the act/360 rate is $5 \times 360/365 = 4.9315\%$, as both must give the same accrued interest on 16 March 2012.

Note Finance often develops extensive shorthand terminology. In the US dollar market, the act/360 daycount convention originally evolved from the money markets, where short-term loans were traded. By contrast the 30/360 daycount convention is a product of the bond markets. As a result practitioners use, for example, the jargon 'annual-money' instead of 'annual compounding with actual three-sixty daycount', and 'semi-bond' instead of 'semi-annual compounding with thirty three-sixty daycount'. Newcomers to a tradefloor often reveal themselves as rookies by using the longhand version and being unaware of the industry jargon—as I did in my first months on the job.

Note Interest rate derivatives often have quarterly payment dates in the middle of the month since many contracts are based on ***IMM dates*** (named after the International Monetary Market where the date system originated), the third Wednesday of March, June, September and December. We encounter IMM dates when we discuss interest rate futures in Chapter 5.

1.5 An abridged guide to stocks, bonds and FX

This book concerns derivative contracts whose value derives from the value of an underlying asset. Here we briefly define the most common such assets: stocks, bonds and foreign exchange. Derivative valuation usually takes the current price of the underlying asset as a given, and so this section is notably and deliberately brief.

A ***stock*** or ***share*** is an asset giving ownership in a fraction of a company. A stock may at times pay a dividend, a cash payment often expressed as a percentage q of the stock price, usually related to the profitability of the company. The stock is publicly traded and its price easily observed. We denote the price at time t of a stock by S_t. The known current price is called the ***spot*** price. We do not need to know much more about stocks in this book.

A ***fixed rate bond*** with coupon c and notional N is an asset that pays a coupon cN each year (or at some other frequency) and notional N back at its maturity date T. Note that a bond of notional 1 is an annuity of c plus a zero coupon bond with maturity T. We also encounter ***floating rate bonds*** where the coupon is linked to a variable interest rate. These are defined in Chapter 4.

The term ***security*** encompasses stocks and bonds, and broadly covers traded, negotiable instruments that have financial value. An ***asset*** is a broader definition encompassing securities and instruments that may not be readily tradable such as real estate, although often in finance, asset and security are conflated.

Note Student financial aid is an asset for the student, but is not a security as it cannot be bought or sold readily. Similarly, royalties for the David Bowie catalogue are an asset. However, assets may sometimes be ***securitized***. For example, financial engineers in 1997 constructed (and David Bowie sold) a Bowie Bond whose coupons were linked to the royalty stream, which was then traded in the market. Similarly, a homeowner's mortgage is an asset for the lending bank, and there was a large industry central to the financial industry securitizing mortgages by packaging them into a form that could be bought and sold readily. During the financial crisis the value of these securities fell dramatically as underlying house prices fell and the ability of homeowners to continue to pay interest payments on their mortgages declined.

Foreign currency is a holding of cash in a non-dollar currency. Its value in US dollars (USD) will depend on the foreign exchange (FX) rate. Market convention is that euro (symbol € and three-letter abbreviation EUR), pounds sterling (£, GBP), Australian dollars (AUD) and New Zealand dollars (NZD) are usually quoted as the amount of US dollars per unit of foreign currency. All other currencies are usually quoted as foreign currency per one US dollar. The ordering of the first five currencies is set as above, meaning the market convention is usually to quote, for example, GBP per €1 and AUD per £1.

Derivative contracts are entered into between two entities or ***counterparties***. A counterparty can be a bank like Societe Generale, a sovereign country like Greece, a government sponsored entity like Freddie Mac, a local government like Orange County, a hedge fund like Long Term Capital Management, a corporation like Parmalat or Procter and Gamble, or even a high net worth individual.

We have established the basics of discounting and compounding, along with the definitions of the most commonplace assets underlying derivatives. We are now ready to introduce our first derivative contract.

1.6 EXERCISES

1. **Daycount and frequency**

   Two market standards for US dollar interest rates are semi-annual compounding with 30/360 daycount (semi-bond, denoted y_{SB}), and annual compounding with act/360 daycount (annual-money, denoted y_{AM}).

(a) Derive an expression for y_{AM} in terms of y_{SB}. You can assume all years have 365 days.

(b) Show that when $y_{SB} = 6\%$, $| y_{SB} - y_{AM} | < 0.01\%$. Calculate y_{AM} for the cases $y_{SB} = 5\%$ and $y_{SB} = 7\%$ and use your answers to deduce which rate has larger standard deviation.

(c) The ***volatility*** of a rate y is usually defined as the standard deviation of its logarithm, and can be well approximated by $\frac{dy}{y}$. By differentiating your answer to (a), derive an expression for the ratio of the volatility of y_{AM} to the volatility of y_{SB}, assuming that the current level of rates is $y_{SB} = 6\%$. Compare your answer to your empirical result from (b).

2. **Simple interest**
A ***simple interest rate*** of r for T years means a 100 investment becomes $100(1 + rT)$ at maturity T. (In other words, there is no compounding.)

(a) For simple interest of 5% for ten years, calculate the equivalent interest rate with (i) annual, (ii) quarterly and (iii) continuous compounding. Assume 30/360 daycount.

(b) Show that if simple interest of r for T years is equivalent to r^* interest rate with annual compounding, then $r^* \to 0$ as $T \to \infty$.

3. **Non-standard annuity**
Suppose annually compounded zero rates for all maturities (with 30/360 daycount) are r. An annuity pays n at times $n = 1, \ldots, N$.

(a) What is the present value of the annuity?

(b) What is the present value of the infinite annuity as $N \to \infty$?

PART II

Forwards, Swaps and Options

2

Forward contracts and forward prices

2.1 Derivative contracts

A ***derivative contract***, or simply ***derivative***, is a financial contract between two counterparties whose value is a function of—derives from, hence derivative—the value of another variable, for example the price of a security such as a stock.

Note The definition of a derivative contract has nothing to do with calculus or differentiation.

For example, a contract where one counterparty pays another counterparty one dollar in one year's time if the amount of snowfall in Boston over the year is above 50 inches is a weather derivative. In this book we will exclusively consider ***financial derivatives***, derivatives of financial assets like stocks or variables like interest rates.

Derivatives traded directly between two counterparties are called ***over-the-counter (OTC)*** derivatives. Most contracts described later in this book are typically OTC, in particular the interest rate derivatives described in Chapters 4 and 12. OTC derivatives contrast with ***exchange-traded*** derivatives where an exchange matches buyers and sellers and each counterparty faces the exchange on the contract. Exchange-traded contracts are exemplified by futures which we outline in Chapter 5.

Derivatives are often defined by the payout at some specified maturity date. For example, the weather derivative above has payout $g(S_T)$, where

$$g(S_T) = I\{S_T > 50\} = \begin{cases} \$1 & \text{if } S_T > 50 \\ 0 & \text{otherwise,} \end{cases}$$

and S_T is the total snowfall in inches during the year up to $T = 1$. Note that both S_T and $g(S_T)$ are random variables whose value is unknown until $T = 1$.

Derivatives can also be defined by the contractual obligations of the two counterparties involved, and the payout function deduced from these. This is the case with the first financial derivative we encounter, the forward contract.

2.2 Forward contracts

A ***forward contract***, or simply ***forward***, is an agreement between two counterparties to trade a specific asset, for example a stock, at a certain future time T and at a certain price K. At the current time $t \leq T$, one counterparty agrees to buy the asset at T, and is ***long*** the forward contract. The other counterparty agrees to sell the asset, and is ***short*** the forward contract. The specified price K is known as the ***delivery price***. The specified time T is known as the ***maturity***.

A forward contract is easily defined, yet often causes confusion when first encountered. In particular, it is important to understand the distinction between the terms inherent in the contract (the underlying asset, K and T), which are fixed when the contract is agreed upon, and the value of the contract, which will vary over time.

We define $V_K(t, T)$ to be the ***value*** at current time $t \leq T$ of being long a forward contract with delivery price K and maturity T, that is, how much the contract itself is worth at time t to the counterparty which is long.

Since the counterparty long the forward contract must pay K at T to buy an asset which is worth S_T, we immediately have $V_K(T, T) = S_T - K$, and so we have restated the forward contract in terms of the linear payout function $g(S_T) = S_T - K$. Similarly, the payout at maturity from a short forward contract is $K - S_T$.

We use the terms 'value at maturity' and 'payout' interchangeably to describe $V_K(T, T)$, although we discuss later the subtle difference between the forward contract defined above where the underlying asset is actually traded at T, and a derivative contract defined simply to have value $S_T - K$ at T.

We define the ***forward price*** $F(t, T)$ at current time $t \leq T$ to be the delivery price K such that $V_K(t, T) = 0$, that is, such that the forward contract has zero value at time t. In particular, by definition one can at t go long—or 'buy'—a forward contract with delivery price $K = F(t, T)$ for no upfront cost. From above, we immediately have $F(T, T) = S_T$.

$F(t, T)$ is the special delivery price satisfying $V_{F(t,T)}(t, T) = 0$. However, we can at time t enter into a forward contract with any delivery price K, provided we pay $V_K(t, T)$ to do so.

We have not yet determined $V_K(t, T)$ or $F(t, T)$ for general $t < T$, and we spend the following sections investigating these quantities. However, we can establish now the important distinction between $F(t, T)$ and $V_K(t, T)$ with the following simple example. Suppose a stock which pays no dividends always has price 100 and interest rates are always zero. Then you should convince yourself that $F(t, T) = 100$ and $V_K(t, T) = 100 - K$ for all $t \leq T$.

2.3 Forward on asset paying no income

We let r be the constant zero rate with continuous compounding, and for now suppress any dependence of the interest rate on time.

Result *For an asset paying no income, for example a stock that pays no dividends,*

$$F(t, T) = S_t e^{r(T-t)}.$$

When $t = 0$ we have the simple expression

$$F(0, T) = S_0 e^{rT}.$$

Note We can rewrite $F(t, T) = S_t/Z(t, T)$, the ratio of the stock price to the ZCB price. Ratios of this form play an important role in Chapter 9.

We prove this result two ways, introducing important concepts that occur throughout quantitative finance. The first method we term the ***replication proof***, which we saw briefly in Chapter 1. If we can demonstrate that two portfolios (where we define a portfolio simply to be a holding of a linear combination of assets) always have the same value at time T, and we have neither added nor subtracted anything of non-zero value between t and T, then the portfolios must have the same value at time $t \leq T$.

The second method we term the ***no-arbitrage*** or 'no free money' proof. We here define ***arbitrage*** to be a situation where we start with an empty portfolio and simply by executing market transactions end up for sure with a portfolio of positive value at time T. A no-arbitrage proof is based on the assumption that there exist no such situations.

In Chapter 6 we more formally define the concept of arbitrage, and show that these two methods of proof are equivalent. In particular, we show that the assumption of no-arbitrage, appropriately defined, allows us to adopt formally proof by replication.

Proof I **Replication.** At current time t, we let portfolio **A** consist of one unit of stock and portfolio **B** consist of long one forward contract with delivery price K, plus $Ke^{-r(T-t)}$ of cash which we deposit at the interest rate r.

At time T portfolio **A** has value S_T. In portfolio **B** at T we have an amount K of cash, which we use to buy the stock at T via the forward contract. Therefore, portfolio **B** also has value S_T at T. Alternatively, we can think of portfolio **B** at T as K of cash and a forward contract with value $S_T - K$.

Since the value of these portfolios at time T is the same, and we have neither added nor taken away any assets of non-zero value, their value at t is the same. Therefore,

$$S_t = V_K(t, T) + Ke^{-r(T-t)}.$$

The forward price $F(t, T)$ is the value of K such that $V_K(t, T) = 0$, and thus satisfies $S_t = F(t, T)e^{-r(T-t)}$. ■

Proof II **No-arbitrage.** Suppose $F(t, T) > S_t e^{r(T-t)}$. At current time t we execute three transactions. We go short one forward contract ('sell the stock forward') at its forward price $F(t, T)$—remember we can do this at no upfront cost. We borrow S_t cash at interest rate r for time $T - t$, and with the cash we buy the stock at its current market price S_t.

At time T we must sell the stock for $F(t, T)$ under the terms of the forward contract, and we pay back the loan amount of $S_t e^{r(T-t)}$. Hence we obtain an amount of cash at T equal to $F(t, T) - S_t e^{r(T-t)} > 0$. Thus we started with an empty portfolio and ended up with a certain profit of $F(t, T) - S_t e^{r(T-t)} > 0$.

Now suppose $F(t, T) < S_t e^{r(T-t)}$. At time t we can go long one forward contract with delivery price $F(t, T)$ for no cost, sell the stock for S_t, and deposit the proceeds at r for time $T - t$. (You may ask what happens if we do not have the stock to sell. We address this in our review of assumptions.)

At time T we receive $S_t e^{r(T-t)}$ at the maturity of the deposit, and must buy back the stock via the forward contract for $F(t,T)$. Hence we obtain a profit of $S_t e^{r(T-t)} - F(t,T) > 0$.

Therefore, under the assumption of no-arbitrage, we must have $F(t,T) = S_t e^{r(T-t)}$. In all other cases we can construct an arbitrage portfolio. ■

No-arbitrage arguments play an important role throughout quantitative finance. Proof II is our first encounter with such an argument and it is important to get comfortable with its construction. If $F(t,T) \neq S_t e^{r(T-t)}$ one can start with an empty portfolio and end up for sure with a portfolio of positive value. Intuitively, to construct the arbitrage when the forward price $F(t,T)$ is 'too high' relative to the spot stock price S_t, we 'sell' it (that is, sell the stock forward) and buy the stock spot.

Note that the forward price depends only on the current stock price S_t, the interest rate r and the time to maturity $T - t$. It often surprises those encountering forwards for the first time that the determination of the forward price does not depend on the growth rate or standard deviation of the stock, or indeed any distributional assumptions about S_T. The forward price says nothing further about predicting where the stock will be at time T than the spot price does. Furthermore, any two assets which pay no income and which have the same spot price S_t will have the same forward price regardless of any views about their future movements.

To understand this heuristically, suppose one has S_t of cash. One can either buy the stock today, or invest the cash at rate r and agree today to buy the stock forward at time T. The invested cash grows to $S_t e^{r(T-t)}$; thus the forward price has to equal this for one to be indifferent between the two strategies, which both result simply in a holding of one stock at T. Alternatively, consider the differences and similarities between (a) going long a forward contract (at its forward price) and (b) starting with no cash, borrowing cash to buy the stock, selling the stock at time T and repaying the loan at T. You should convince yourself that both portfolios have value zero at time t and value $S_T - S_t e^{r(T-t)}$ at time T.

Trading a forward contract, however, does allow one to gain exposure to movements in the stock price in a capital-efficient way, that is, without initially having to pay out or borrow the purchase price at time t. Forwards allow institutions to establish larger exposures for a given amount of initial capital. Indeed, much of the current regulatory reform is concerned with appropriate capital requirements for institutions entering into forwards and other similar derivatives.

Note Often the stock is itself used as collateral for the loan taken out to purchase it.

2.4 Forward on asset paying known income

Result *Suppose an asset pays a known amount of income (for example, dividends, coupons or even rent) during the life of the forward contract, and the present value at t of the income is I. Then*

$$F(t,T) = (S_t - I)e^{r(T-t)}.$$

Proof I **Replication.** Let portfolio **A** consist of one unit of the asset and $-I$ cash. (Note that a negative amount of cash can be thought of simply as a debt or overdraft.) Let portfolio **B** consist of long one forward contract with delivery price K, plus $Ke^{-r(T-t)}$ cash.

At time T, portfolio **A** has value $S_T + Ie^{r(T-t)} - Ie^{r(T-t)} = S_T$, the second term on the left hand side being the value at T of the income received. Portfolio **B** has again value S_T by the same argument as before. Therefore, the values of the two portfolios at time t are equal, and thus

$$S_t - I = V_K(t, T) + Ke^{-r(T-t)}.$$

Hence $F(t, T) = (S_t - I)e^{r(T-t)}$. ■

Proof II **No-arbitrage.** We assume the equality does not hold, and show that in this case we can create a portfolio with positive value from an initial portfolio of zero value.

Suppose $F(t, T) > (S_t - I)e^{r(T-t)}$. We go short one forward contract, borrow S_t of cash and buy one stock, which provides income with value $Ie^{r(T-t)}$ at T.

At time T we sell the stock for $F(t, T)$ via the forward contract and pay back the $S_t e^{r(T-t)}$ loan. Therefore, we obtain a profit of $F(t, T) - (S_t - I)e^{r(T-t)} > 0$.

A similar argument applies if $F(t, T) < (S_t - I)e^{r(T-t)}$. ■

2.5 Review of assumptions

Let us pause to consider assumptions we have either implicitly or explicitly made to construct our arguments, and which underpin much of the logic of quantitative finance. The extent to which these assumptions are valid in practice is a key determining factor of the nature of financial markets.

1 We can borrow or lend money freely at rate r. In practice, there will be an offered rate r_{OFF} at which one borrows and a bid rate r_{BID} at which one lends, $r_{BID} \leq r_{OFF}$. Arbitrage arguments then give for the non-dividend paying stock

$$S_t e^{r_{BID}(T-t)} \leq F(t, T) \leq S_t e^{r_{OFF}(T-t)}.$$

However, for large market participants, much borrowing and lending nets and is effectively done close to one rate r.

2 We can always buy or sell as much of the asset as we want, and the market is sufficiently deep that the amount we buy or sell does not move the price. This usually holds reasonably well for foreign exchange and interest rate swap markets. We also assume we can buy or sell assets with negligible transaction costs. In some markets—for example, US treasury bonds, bond futures and foreign exchange—bid and offer prices are very close together, and transaction costs are indeed low.

3 We can ***sell assets short*** at will, that is, have negative amounts of assets we do not own. This often represents the largest conceptual hurdle. For ease of understanding, we can simply assume we always have the asset we want to sell, or that there is another market participant who holds the asset and can sell it, and for whom the same no-arbitrage arguments apply. In practice, we can usually sell a stock short, without owning it. To do so, we borrow the stock from someone who owns it (for example, an asset manager or broker); sell it; buy it back at later time; and then return the stock to the asset manager. The owner of the stock keeps the rights to any dividends during the life of the short sale.

   Whilst the mechanics of a short sale are important in practice, mathematically one can understand short selling best simply by assuming we can hold quantity λ of an asset, for all $\lambda \in \mathbb{R}$. In reality this may or may not be the case depending on the ability to obtain a stock to borrow, and the regulations surrounding short selling.

4 People always want to exploit arbitrages in order to obtain 'free money', and have the capacity to do so. In other words, arbitrages cannot persist in the market and arbitrageurs are always present.

Many of these assumptions can be questioned and challenged. Indeed, several can be violated to some degree even in normal market conditions. A noteworthy feature of the market turmoil in 2008-2009 was gross violations of many of them, viz:

1 Money markets froze as financial market participants were gripped by fear of counterparty risk and bankruptcy. It became very hard to borrow money on an unsecured basis as no one wanted to lend in an environment where major institutions could go bankrupt. The experience of a German bank lending to Lehman Brothers, the US investment bank, in September 2008 was salutary. Published levels for interest rates bore little relation to the actual rates banks were charging for loans. Question 4 in the exercises explores this.

2 Liquidity evaporated, even in usually deep markets. The sizes one could execute became small, and the effect of any trade executed became outsized. Transaction costs increased as liquidity decreased and as market makers found it far harder to get out of any positions. A general reduction in risk appetite led to wholesale liquidation of positions which themselves caused market prices to move dramatically and discontinuously.

3 Regulators re-examined short selling rules, causing further market turmoil. In certain jurisdictions, short selling was banned for particular assets or for particular time periods. In addition, the ability to find stock to borrow, given heightened concern about counterparty bankruptcy, was significantly reduced.

4 The least attention was probably paid by market participants to the fourth assumption, that arbitrageurs exist and that they take advantage of opportunities. In the financial market turbulence that followed the bankruptcy of Lehman Brothers in September 2008 there was sufficient fear within financial markets, along with

reduced appetite to take risk, execute trades or suffer even temporary mark-to-market losses, that broad arbitrages existed for remarkably long periods. We encounter some such examples in Chapters 6 and 13.

2.6 Value of forward contract

Recall from our example of the asset that pays no income, that the replication argument gave

$$V_K(t,T) + Ke^{-r(T-t)} = S_t.$$

If we substitute $F(t,T) = S_t e^{r(T-t)}$ we obtain $V_K(t,T) = (F(t,T) - K)e^{-r(T-t)}$, the difference between the forward price and delivery price discounted back to today. We now show this result holds in general for all assets.

Result *The value of the forward contract on an asset satisfies*

$$V_K(t,T) = (F(t,T) - K)e^{-r(T-t)}.$$

Proof Suppose $V_K(t,T) > (F(t,T) - K)e^{-r(T-t)}$.

At time t we go long a forward contract with delivery price $F(t,T)$ (at no cost), and go short a forward contract with delivery price K. For the latter we receive $V_K(t,T)$ at t, which is invested at rate r.

At maturity T, the payout of the two forward contracts is

$$(S_T - F(t,T)) + (K - S_T) = -(F(t,T) - K).$$

Therefore, the value of portfolio at time T is $V_K(t,T)e^{r(T-t)} - (F(t,T) - K) > 0$.

A similar argument holds if $V_K(t,T) < (F(t,T) - K)e^{-r(T-t)}$. We go long a forward contract with delivery price K, paying $V_K(t,T)$ at t to do so, and short a forward contract with delivery price $F(t,T)$ for no cost. ■

Fixing a specific time t_0, note the distinction between:

$$V_{F(t_0,T)}(t,T), t_0 \leq t \leq T,$$

which is zero when $t = t_0$, then equals the value of a specific forward contract over time; and

$$V_{F(t,T)}(t,T), t_0 \leq t \leq T,$$

which is identically zero.

2.7 Forward on stock paying dividends and on currency

Result *Suppose a stock pays dividends at a known dividend yield q, expressed as a percentage of the stock price on a continually compounded per annum basis. Then the forward price satisfies*

$$F(t,T) = S_t e^{(r-q)(T-t)}.$$

Proof Let portfolio **A** be $e^{-q(T-t)}$ units of the stock, with dividends all being reinvested in the stock, and let portfolio **B** again be one forward contract with delivery price K plus $Ke^{-r(T-t)}$ of cash.

At time T, both portfolios have value S_T. Therefore, their values at current time t are the same and thus $S_t e^{-q(T-t)} = V_K(t,T) + Ke^{-r(T-t)}$. We again obtain the result by setting $V_K(t,T) = 0$. ■

Similarly to the example where the asset pays a known income, the presence of dividends lowers the forward price. There is an advantage to buying the asset spot versus buying it forward, since in the latter case we do not receive any dividends paid before T. The forward price is lower in order to compensate. The forward contract on a foreign currency is similar.

Result *Suppose X_t is the price at time t in dollars of one unit of foreign currency. (For example, at the time of writing £1 = \$1.55 and €1 = \$1.31.) Let $r_\$$ be the dollar zero rate and r_f the foreign zero rate, both constant and continuously compounded. Then the forward price satisfies*

$$F(t,T) = X_t e^{(r_\$ - r_f)(T-t)}.$$

Proof I Simply replace q by r_f in the result for a stock paying dividends. The foreign currency is analogous to an asset paying a known dividend yield, the foreign interest rate. ■

Proof II **Replication.** See Question 3 in the exercises. ■

2.8 Physical versus cash settlement

We have assumed forwards are ***physically settled***, meaning that one actually pays K and receives the asset at time T. However, some forwards are ***cash settled***, meaning one simply receives (pays if negative) the amount $S_T - K$ at T. We encounter both cash and physically settled contracts later in the book. For example, the derivative contract called an interest rate swaption (Chapter 12) can upon execution of the contract be specified as physical or cash settlement. Often forwards on financial assets are cash settled when the price S_T is well defined and fixes on a screen. Cash settled forwards are also sometimes known as ***contracts for difference***. Both have the same value at T, and hence for $t \leq T$. However, a cash settled forward has no further exposure to the asset price, whilst a physically settled contract—where one owns the asset at T—continues to have exposure to asset price movements. Thus cash and physical settlement are different in their risk exposure after T.

Cash-settled forwards have many similarities with ***spread bets***, and indeed many spread betting companies are populated by former derivative traders. For example, if I 'buy' Boston Red Sox regular season wins at 90, in size one dollar per win, I have a long forward position with delivery price 90. If the Red Sox win 100 games I make \$10. If they have 81 wins I lose \$9. The payout is $S_T - K$ where S_T is number of wins. A key difference is that S_T is not the price of a security or tradable asset and so the replication and no-arbitrage arguments for forward pricing do not apply. The price of the spread bet must come from supply and demand, these presumably being generated by some view of how good the Red Sox are.

2.9 Summary

We have in this chapter derived forward prices for the underlying assets shown in Table 2.1.

Foreign exchange forward contracts are widely traded and highly liquid, with approximately \$60 trillion notional of outstanding contracts as of December 2011. They are employed extensively by a range of institutions to manage foreign exchange exposure of future cashflows. Forward contracts on bonds, stocks and stock indices are less prevalent with, for example, only \$2 trillion of outstanding equity-linked forward contracts as of December 2011. However, futures contracts—a variant of forwards—on bonds and stock indices are highly liquid with hundreds of thousands of contracts traded each day. We detail futures contracts in Chapter 5. The largest derivative market by far is the interest rate derivative market. In the next two chapters we introduce forward rate agreements and interest rate swaps, which constitute the vast majority of the \$500 trillion notional of outstanding interest rate contracts.

Table 2.1 Forward prices for various assets.

Underlying	**Forward price**
Asset paying no income	$S_t e^{r(T-t)}$
Asset paying known income I	$(S_t - I)e^{r(T-t)}$
Asset paying dividends at rate q	$S_t e^{(r-q)(T-t)}$
Foreign exchange	$X_t e^{(r_\$ - r_f)(T-t)}$

2.10 EXERCISES

1. **Forwards in presence of bid-offer spreads**
   Let S_t be the current price of a stock that pays no dividends.
   (a) Let r_{BID} be the interest rate at which one can invest/lend money, and r_{OFF} be the interest rate at which one can borrow money, $r_{BID} \leq r_{OFF}$. Both rates are continuously compounded. Using arbitrage arguments, find upper and lower bounds for the forward price of the stock for a forward contract with maturity $T > t$.
   (b) How does your answer change if the stock itself has bid price $S_{t,BID}$ and offer price $S_{t,OFF}$?
2. **Forwards and arbitrage**
   At time t you own one stock that pays no dividends, and observe that

$$F(t, T) < \frac{S_t}{Z(t, T)}.$$

   What arbitrage is available to you, assuming you can only trade the stock, ZCB and forward contract? Be precise about the transactions you should execute to exploit the arbitrage.
3. **FX forwards**
   By constructing two different portfolios, both of which are worth one unit of foreign currency at time T, prove by replication that the forward price at time t for one unit of foreign currency is given by

$$F(t, T) = X_t e^{(r_\$ - r_f)(T-t)},$$

   where X_t is the price at time t of one unit of foreign currency and T is the maturity of the forward contract.
4. **FX forwards during the financial crisis**
   FX forwards are among the most liquid derivative contracts in the world and often reveal more about the health of money markets (markets for borrowing or lending cash) than published short-term interest rates themselves.
   (a) On 3 October 2008, the euro dollar FX rate was trading at €1 = \$1.3772, and the forward price for a 3 April 2009 forward contract was \$1.3891. Assuming six-month euro interest rates were 5.415%, what is the implied six-month dollar rate? Both interest rates are quoted with act/360 daycount and semi-annual compounding. There are 182 days between 3 October 2008 and 3 April 2009.
   (b) Published six-month dollar rates were actually 4.13125%. What arbitrage opportunity existed? What does a potential arbitrageur need to transact to exploit this opportunity?

(c) During the financial crisis, several European commercial banks badly needed to borrow dollar cash, but their only source of funds was euro cash from the European Central Bank (ECB). These banks would: borrow euro cash for six months from the ECB; sell euros/buy dollars in the spot FX market; and sell dollars/buy euros six months forward (to neutralize the FX risk on their euro liability). Explain briefly how these actions may have created the arbitrage opportunity in (b), which existed for several months in late 2008.

5. **Forwards and carry**

(a) Use arbitrage arguments involving two forward contracts with maturity T to prove that

$$V_K(t,T) = (F(t,T) - K)\, e^{-r(T-t)}.$$

(b) Verify that $V_K(T,T)$ equals the payout of a forward contract with delivery price K. For an asset that pays no income, substitute the expression for its forward price into the above equation and give an intuitive explanation for the resulting expression.

(c) Suppose at time t_0 you go short a forward contract with maturity T (and with delivery price equal to the forward price). At time t, $t_0 < t < T$, suppose both the price of the asset and interest rates are unchanged. How much money have you made or lost? (This is sometimes called the ***carry*** of the trade.) How does your answer change if the asset pays dividends at constant rate q?

6. **Dollar-yen and the carry trade**

A major currency pair is dollar-yen, quoted in yen per dollar. Suppose the current price is \$1 = ¥82.10. Suppose also that the five-year dollar interest rate is 2.17% and the two-year rate is 0.78% (semi-annual, 30/360 daycount), and that the five-year and two-year yen rates are 0.63% and 0.41% respectively (semi-annual, act/365 daycount).

(a) Calculate the forward price for dollar-yen five years forward. For simplicity use a 0.5 accrual factor, rather than 182/365 etc, for yen.

(b) Suppose you were unable to trade forward contracts, but were able to trade spot FX and borrow or lend dollar and yen cash. How could you synthetically go long the forward contract in (a)?

(c) Suppose you went long one forward contract from (a), and three years from now the FX rate and interest rates are unchanged. What is your profit or loss on the forward?

7. **FX forward brainteaser, part 1**

Let X_t be the euro dollar exchange rate (in dollars per euro) at time t, and suppose $X_T \mid X_t \sim \text{lognormal}(\log X_t + \mu(T-t), \sigma^2(T-t))$.

(a) Calculate $E(X_T \mid X_t)$. Find in terms of $r_\$$, $r_€$ and σ the value of μ such that $E(X_T \mid X_t) = F(t,T)$, where $F(t,T)$ is the forward euro dollar FX rate (see Question 3).

(b) Let Y_t be the dollar euro exchange rate (in euro per dollar). Find the distribution of $Y_T \mid Y_t$ and calculate $E(Y_T \mid Y_t)$. (**Hint** Lengthy calculations are not required.) For what value of μ does $E(Y_T \mid Y_t) = \tilde{F}(t,T)$, where $\tilde{F}(t,T)$ is the forward dollar euro FX rate?

(c) What condition has to be satisfied such that both $E(X_T \mid X_t) = F(t,T)$ and $E(Y_T \mid Y_t) = \tilde{F}(t,T)$? What does this imply in practice?

(d) Suppose that $X_t = 1$. Show that for the value of μ found in (a), $E(X_T \mid X_t = 1) = E(Y_T \mid Y_t = 1)$ if and only if $\sigma = \sqrt{2(r_\$ - r_€)}$ and $r_\$ \geq r_€$.

Note Does the lack of symmetry in results (c) and (d) trouble you?

8. **Real estate forwards**

A house in Boston is offered for sale at $1 million. Interest-only mortgage rates are 4% (annual compounding) and the house can be rented out for $5,000 per month. Real estate taxes to be paid by the owner are $10,000 per annum.

(a) Find an upper bound for the one-year forward price for the house. State any assumptions you are making.

(b) Can you establish a lower bound on the one-year forward price? Consider briefly the cases where:

(i) you own the house and live in it

(ii) you own the house and rent it out

(iii) you own a house in Boston and are prepared to move

(iv) you own no real estate.

(c) There is in fact a nascent market in real estate forwards, largely for institutional real estate investors. At the height of the financial crisis, the difference between the theoretical upper bound and the actual forward price increased significantly. Discuss briefly why this might have occurred.

3

Forward rates and libor

We now apply our forward contract machinery to the case where the underlying asset is a zero coupon bond (ZCB). This construct leads naturally to the concept of forward rates.

3.1 Forward zero coupon bond prices

For $T_1 \leq T_2$, consider a forward contract with maturity T_1 on a ZCB with maturity T_2. That is, the underlying security price is $Z(t, T_2)$, the price at t of the T_2-maturity ZCB, rather than S_t.

Consider the forward price of this forward contract, denoted $F(t, T_1, T_2)$, the delivery price such that the forward contract has zero value at time t. Equivalently, the forward price is where one can (for no upfront cost) agree at t to buy at T_1 the ZCB with maturity T_2.

Result

$$F(t, T_1, T_2) = \frac{Z(t, T_2)}{Z(t, T_1)}.$$

Proof Let portfolio **A** be one ZCB maturing at T_2 and portfolio **B** be one forward contract with delivery price K, and K ZCBs maturing at T_1.

Portfolio **A** is worth 1 at time T_2 by definition. In portfolio **B**, the holding of the K ZCBs with maturity T_1 is worth K at time T_1. One can use this to buy a ZCB with maturity T_2 via the forward contract, and thus portfolio **B** is also worth 1 at time T_2. Therefore, both portfolios must be worth the same at time t and so

$$Z(t, T_2) = V_K(t, T_1) + KZ(t, T_1).$$

The forward price is the value of K such that $V_K(t, T_1) = 0$. Therefore,

$$F(t, T_1, T_2) = \frac{Z(t, T_2)}{Z(t, T_1)}.$$

■

3.2 Forward interest rates

The ***forward rate*** at current time t for period T_1 to T_2, $t \leq T_1 \leq T_2$, is the rate agreed at t at which one can borrow or lend money from T_1 to T_2.

In this section we use the simplified notation f_{12} for the forward rate, and we suppose that r_1 and r_2 are the current zero rates for terms T_1 and T_2 respectively.

A graphical representation is the best way to picture forward rates. As shown in Figure 3.1, we can agree at t to lend until T_1 at rate r_1, and to lend for period T_1 to T_2 at rate f_{12}. Alternatively, we can agree at t to lend until T_2 at rate r_2.

A replication argument concludes that the amount accrued at T_2 under the two alternatives must be the same. Thus in the case rates are continuously compounded, the forward rate satisfies

$$e^{r_1(T_1-t)}e^{f_{12}(T_2-T_1)} = e^{r_2(T_2-t)}$$

$$\Rightarrow f_{12} = \frac{r_2(T_2-t) - r_1(T_1-t)}{T_2 - T_1},$$

a weighted average of zero rates.

Similarly, in the case that rates are annually compounded we have

$$(1+r_1)^{(T_1-t)}(1+f_{12})^{(T_2-T_1)} = (1+r_2)^{(T_2-t)}.$$

Since

$$Z(t,T_i) = \frac{1}{(1+r_i)^{(T_i-t)}}, \text{ for } i = 1,2,$$

we see that the forward ZCB price is related to the forward rate by

$$F(t,T_1,T_2) = \frac{1}{(1+f_{12})^{(T_2-T_1)}}.$$

We used simplified notation in this section, but notation for interest rates soon gets unwieldy. In general we have to incorporate the current time, the start date of the interest period, the end date of the interest period, plus the compounding frequency of the rate. Our notation r_1, r_2 and f_{12} suppresses the current time and the compounding frequency, and soon lacks the capacity to distinguish between rates. We introduce more comprehensive notation when we define swaps in Chapter 4.

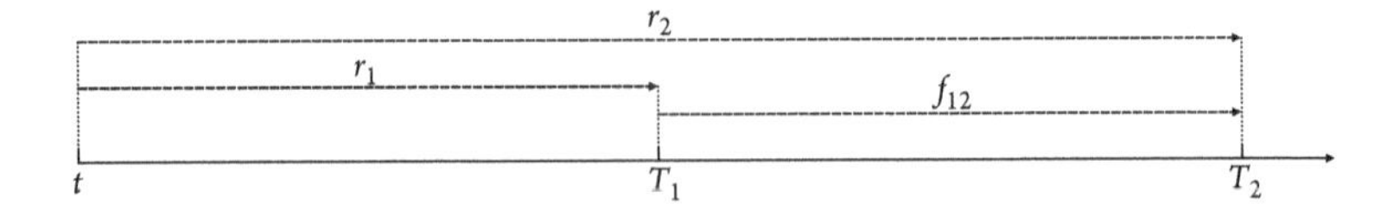

Figure 3.1 Forward and zero rates

3.3 Libor

The rate at which banks borrow or lend to each other is called ***libor***, an acronym for the London InterBank Offered Rate. The vast majority of interest rate derivatives have payouts that are functions of libor rates, short-dated interest rates of fixed term. (The finance industry often refers to 'libor rates' despite the tautology.) On the current day t, libor rates for periods $\alpha = 1$ (twelve-month libor, often abbreviated to 12mL), $\alpha = 0.5$ (six-month libor, 6mL), $\alpha = 0.25$ (three-month libor, 3mL) etc are published. We denote the libor rate at current time t for period t to $t + \alpha$ by $L_t[t, t + \alpha]$, replacing our earlier interest rate notation r. The libor rate $L_T[T, T + \alpha]$ for a future date $T > t$ is a random variable.

Banks can deposit (or borrow) N at time t and receive (or pay back) $N(1+\alpha L_t[t, t+\alpha])$ at time $t + \alpha$. All interest is paid at the maturity or term of the deposit, and there is no interim compounding. Most interest rate derivatives typically reference three-month or six-month libor. Again, we usually assume that the notional $N = 1$ unless otherwise stated.

Note Libor was for many years an obscure financial term (often mispronounced by broadcasters with a short 'i' as in 'liberation' rather than 'lie-bore' which is universal on Wall Street). The aftermath of the financial crisis of 2008–2009, however, shone a spotlight on libor and its setting mechanism. Each libor fix is calculated by polling a panel of banks for their rates, discarding the highest and lowest quartiles and averaging the remainder. Although in the past, observable interbank certificate of deposit rates were largely in line with libor submissions, the rates bank submit do not necessarily have to reflect actual transactions. This framework led to two potential abuses. First, in times of stress, banks could submit artificially low levels so as not to alarm markets by revealing their funding difficulty. Secondly, swap traders, whose underlying derivative contracts we see in Chapter 4 depend heavily on libor levels, could attempt to influence, to their benefit, rate submissions from their own bank. Both phenomena were components of the libor scandal that broke in 2012. In Chapters 4 and 12 we see that libor is integral to the majority of interest rate derivatives.

3.4 Forward rate agreements and forward libor

A ***forward rate agreement*** (FRA—either pronounced 'frah' or read 'f.r.a.') is a forward contract to exchange two cashflows. Specifically, the buyer of the FRA with maturity T and delivery price or fixed rate K agrees at $t \leq T$ to

$$\left.\begin{array}{ll} \text{pay} & \alpha K \\ \text{receive} & \alpha L_T[T, T+\alpha] \end{array}\right\} \text{ at time } T + \alpha.$$

Thus the payout of the FRA is $\alpha(L_T[T, T + \alpha] - K)$ at time $T + \alpha$.

Note We see in Chapter 14 that simply changing the payment date of both legs from $T + \alpha$ to T opens up an unexpected dimension of complexity.

The random variable here is an interest rate, and the FRA looks much like a forward contract on libor, as it appears similar to paying K to 'buy' the libor rate $L_T[T, T + \alpha]$. However, $L_T[T, T + \alpha]$ is not the price of an asset that one can buy or sell like a stock. The only thing related to libor we can trade is the deposit.

The ***forward libor rate***, denoted by $L_t[T, T + \alpha]$, is the value of K such that the FRA has zero value at time $t \leq T$. Note the analogue with the forward price $F(t, T)$ for a forward contract on an asset.

Result *The forward libor rate is given by*

$$L_t[T, T + \alpha] = \frac{Z(t, T) - Z(t, T + \alpha)}{\alpha Z(t, T + \alpha)}.$$

Proof Consider a portfolio consisting of long one ZCB maturing at T and short $(1 + \alpha K)$ ZCBs maturing at $(T + \alpha)$.

At time T we place the 1 from the T-maturity bond in a deposit with interest rate $L_T[T, T + \alpha]$. We are able to do this (at no cost) by definition of $L_T[T, T + \alpha]$.

Therefore, at time $(T + \alpha)$, the portfolio has value

$$(1 + \alpha L_T[T, T + \alpha]) - (1 + \alpha K) = \alpha(L_T[T, T + \alpha] - K),$$

the payout of a FRA. Therefore, the value of the FRA at time t, which we denote $V_K(t, T)$, is given by the value of this portfolio at t, and so we have

$$V_K(t, T) = Z(t, T) - (1 + \alpha K)Z(t, T + \alpha).$$

The forward libor rate $L_t[T, T + \alpha]$ is the value of K such that the FRA has zero value. Hence

$$L_t[T, T + \alpha] = \frac{Z(t, T) - Z(t, T + \alpha)}{\alpha Z(t, T + \alpha)}.$$ ■

Note Rearranging we have

$$Z(t, T + \alpha) = Z(t, T)\frac{1}{1 + \alpha L_t[T, T + \alpha]}.$$

This is analogous to the expression deriving forward rates from zero rates. We represent this expression graphically in Figure 3.2, considering the present value at t of 1 paid at $T + \alpha$.

Note Our notation for the forward libor rate $L_t[T, T + \alpha]$ and the libor fix $L_T[T, T + \alpha]$ are consistent with each other at $t = T$, since at T the FRA has zero value when $K = L_T[T, T + \alpha]$. That is, the forward libor rate at T for period T to $T + \alpha$ is trivially the libor fix at T.

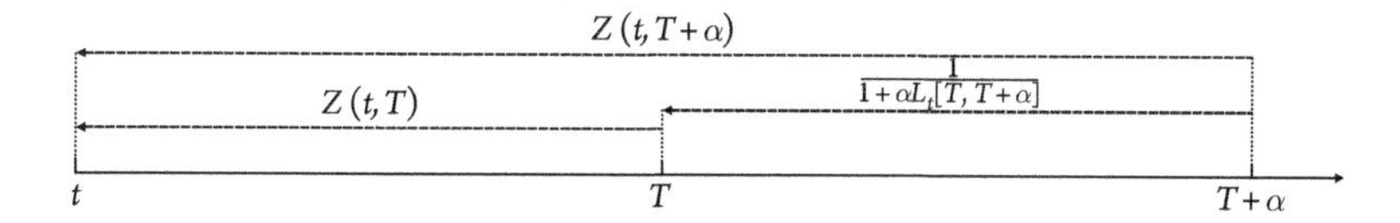

Figure 3.2 Discounting by ZCBs and the forward libor rate

Setting $t = T$ we have

$$Z(T, T + \alpha) = \frac{1}{1 + \alpha L_T[T, T + \alpha]},$$

a familiar expression linking the ZCB price (or discount factor) to the interest rate for that period.

It is important to distinguish between $L_t[t, t + \alpha]$, the libor fix at current time t for deposits from t to $t + \alpha$, and $L_t[T, T + \alpha]$, the forward libor at current time t for period T to $T + \alpha$.

3.5 Valuing floating and fixed cashflows

By definition of forward libor, the value at t of an agreement to receive the unknown floating libor payment $\alpha L_T[T, T + \alpha]$ at $T + \alpha$ equals the value at t of receiving the known quantity $\alpha L_t[T, T + \alpha]$ at $T + \alpha$, which is

$$\alpha L_t[T, T + \alpha]Z(t, T + \alpha) = Z(t, T) - Z(t, T + \alpha),$$

the difference between two ZCB prices.

We thus have established the result that receiving an unknown libor interest payment from T to $T + \alpha$ on a unit of cash has the same value as receiving the unit of cash at T then paying the unit back at $T + \alpha$. This is intuitively clear, since one can invest the unit of cash in a libor deposit, receiving the libor interest payment.

However, the result is non-trivial, in that we have shown that the value at t of agreeing to receive the random quantity $L_T[T, T + \alpha]$ at time $T + \alpha$ is a deterministic function of the known quantities $Z(t, T)$ and $Z(t, T + \alpha)$. The value does not depend on any distributional assumptions about the random variable $L_T[T, T + \alpha]$. This is a key characteristic of forward contracts.

Valuing a fixed cashflow is easy. We know that receiving a fixed payment of K at time $T + \alpha$ has value at time t equal to $KZ(t, T + \alpha)$. We now have all the machinery required to value interest rate swaps.

3.6 EXERCISES

1. **Forwards on zero coupon bonds**
   Suppose $t \leq T_1 \leq T_2 \leq T_3$, where t is current time, and $\Delta > 0$. Recall that $Z(T_1, T_2)$ is the price at time T_1 of a ZCB with maturity T_2, and $F(T_1, T_2, T_3)$ is the forward price at time T_1 for a forward contract with maturity T_2 on a ZCB with maturity T_3.

   (a) For each of the pairs of A and B in Table 3.1, choose the most appropriate relationship out of $\geq, \leq, =$ and ?, where ? means the answer is indeterminate. Give brief reasoning.

Table 3.1 Relationship of A and B.

	A	$\geq, \leq, =$ or ?	B
(i)	$Z(t, T_1)$		1
(ii)	$Z(T_1, T_1)$		1
(iii)	$Z(t, T_2)$		$Z(t, T_3)$
(iv)	$Z(T_1, T_2)$		$Z(T_1, T_3)$
(v)	$Z(T_1, T_3)$		$Z(T_2, T_3)$
(vi)	$Z(T_1, T_1 + \Delta)$		$Z(T_2, T_2 + \Delta)$
(vii)	$F(t, T_1, T_2)$		$F(t, T_1, T_3)$
(viii)	$F(t, T_1, T_3)$		$F(t, T_2, T_3)$
(ix)	$\lim_{T\to\infty} Z(t, T)$		0

   Hint Remember that at current time $t < T$, $F(t, \cdot, \cdot)$ is known, but $Z(T, \cdot)$ is a random variable.

   (b) What can you say about interest rates between T_1 and T_2 if
      (i) $Z(t, T_1) = Z(t, T_2)$?
      (ii) $Z(t, T_1) > 0$ and $Z(t, T_2) = 0$?

2. **Forward rates**
   (a) The one-year and two-year zero rates are 1% and 2% respectively. What is the one-year forward one-year rate (that is, f_{11})? Assume all rates are annually compounded.

   Note This is a favourite question for interviewers.

(b) If the two-year forward one-year rate (f_{21}) is 3%, what is the three-year zero rate?

3. **FRAs**
A bank has borrowing needs at time $T > 0$. Show that by combining an FRA trade today with a libor loan at time T, the bank can today lock in its interest cost for the period T to $T + \alpha$. Does the borrowing bank need to buy or sell the FRA to do this? What is the fixed rate that the bank locks in?

4. **Floating rate annuity**

(a) A derivative contract pays $\alpha L_T[T, T + \alpha]$ at time $T + \alpha$. By constructing a portfolio of ZCBs and a libor deposit that replicates the payout, prove that the value at $t \leq T$ of the derivative contract is $Z(t, T) - Z(t, T + \alpha)$.

(b) Let $T_0, \ldots, T_n$ be a sequence of times, with $T_{i+1} = T_i + \alpha$ for a constant $\alpha > 0$. Use your result from (a) to show that a floating leg of libor payments $\alpha L_{T_i}[T_i, T_i + \alpha]$ at times T_{i+1}, $i = 0, \ldots, n - 1$, has value at time $t \leq T_0$ equal to a simple linear combination of ZCB prices.

(c) Hence find the value of a spot-starting infinite stream of libor payments, that is, when $t = T_0 = 0$ and as $n \to \infty$.

5. **Fixed rate annuity revisited**
Suppose annually compounded zero rates for all maturities are a constant r, so $Z(0, j) = \frac{1}{(1+r)^j}$.

(a) What is the value today of a fixed annuity that pays 1 each year from $T_1 = 1$ to $T_n = n$?

(b) Find the value of the infinite fixed annuity as $n \to \infty$.

6. **Interest rate delta of annuities**
The ***interest rate delta*** of a derivative contract is defined as the partial derivative, $\frac{\partial}{\partial r}$, of its value, and measures sensitivity of the price to interest rate movements. Assume again that $Z(0, j) = \frac{1}{(1+r)^j}$.

(a) Using your results from Questions 4 and 5, calculate the interest rate delta of

(i) a spot-starting annual libor stream ($\alpha = 1$) until time $T = n$.

(ii) a spot-starting fixed rate annuity paying c each year until time $T = n$.

(b) Show that the deltas are equal in magnitude (but opposite in sign) when

$$n = \frac{c(1 + r)}{r(c + r)}((1 + r)^n - 1).$$

(This formula is sometimes referred to as the ***Third Wrangler*** result.)

Verify that when $r = c = 0.06$, the delta of the floating rate annuity is equal in magnitude (but opposite in sign) to that of the fixed rate annuity when $n = 20.734$ years.

(c) Calculate the interest rate delta of (i) an infinite libor stream and (ii) an infinite fixed annuity. Explain your answers.

7. **Floating rate bond**
Let $T_0, \ldots, T_n$ be a sequence of times with $T_{i+1} = T_i + \alpha$ for a constant $\alpha > 0$. A ***floating rate bond*** with notional 1, start date T_0 and maturity T_n pays libor coupons of $\alpha L_{T_i}[T_i, T_i + \alpha]$ at times T_{i+1}, for $i = 0, \ldots, n-1$, and notional 1 at T_n.
   (a) Find the price at $t < T_0$ of the floating rate bond.
   (b) Using a replication argument, find the forward price at t for the floating rate bond (for a forward contract with maturity T), $t < T < T_0$.

4

Interest rate swaps

Interest rate swaps are the most widely traded, liquid and universal of all over-the-counter derivative contracts. The first swap was traded in 1981 and explosive growth followed in the 1990s and 2000s. The outstanding notional of interest rate swaps grew from approximately \$20 trillion in 1995 to \$55 trillion in 2000 to \$400 trillion as of December 2011 (out of a total \$500 trillion interest rate derivative market). The notional numbers are huge compared to US GDP of approximately \$15 trillion and total worldwide equity market capitalization of approximately \$50 trillion. However, as mentioned in Chapter 1, the values of interest rate derivatives are typically small fractions of the notional.

Swaps allow institutions to manage their exposure to interest rate movements or to adjust the nature of their interest rate liabilities. Swaps can be a trading tool, in particular allowing a counterparty to 'buy' or 'sell' interest rates and thus profit (or lose) on their movements, or a risk management tool, for example, allowing a counterparty to replace unknown floating cashflows with a fixed stream.

An excellent and comprehensive summary of interest rate swaps and related derivatives is given by my former colleague at Morgan Stanley, Howard Corb (Corb, 2012).

Note The vast majority of interest rate swaps are over-the-counter contracts. However, the Dodd-Frank Wall Street Reform and Consumer Protection Act, signed into law in 2010, requires—among many other things—that standard interest rate swaps should be cleared through an exchange, once appropriate rules have been established.

4.1 Swap definition

A ***swap*** is an agreement between two counterparties to exchange a series of cashflows at agreed dates. Cashflows are calculated on a notional amount, which again we will typically take to be 1. A swap has a start date T_0, maturity T_n, and payment dates T_i, for $i = 1, \ldots, n$.

The payment frequency for the floating and fixed legs may differ, but we will assume here they are the same. Question 4 in the exercises shows that the frequency of the floating leg is irrelevant for theoretical valuation.

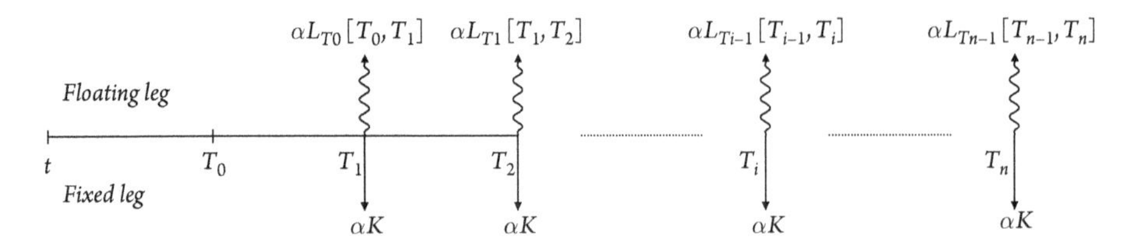

Figure 4.1 A swap

In a standard or ***vanilla*** swap we typically have $T_{i+1} = T_i + \alpha$ for a fixed α. In practice, α may differ slightly for each period (for example, 92/365 or 91/365 instead of 0.25) and should precisely be written α_i, but we will assume constant α for simplicity.

The ***floating leg*** of the swap consists of payments $\alpha L_{T_i}[T_i, T_i + \alpha]$ at $T_i + \alpha$, that is, libor fixing at T_i for the period T_i to $T_i + \alpha$, paid at $T_i + \alpha$.

The ***fixed leg*** of the swap consists of payments αK at $T_i + \alpha$, that is, a fixed rate K accrued from T_i to $T_i + \alpha$, paid at $T_i + \alpha$.

A swap is shown graphically in Figure 4.1. One counterparty (the 'payer') pays the fixed leg to, and receives the floating leg from, the other counterparty (the 'receiver').

4.2 Forward swap rate and swap value

The value of the fixed leg is given by

$$V_K^{FXD}(t) = K \sum_{i=1}^{n} \alpha Z(t, T_i) = KP_t[T_0, T_n]$$

where we define

$$P_t[T_0, T_n] = \sum_{i=1}^{n} \alpha Z(t, T_i).$$

The fixed leg of the swap has the present value of an annuity paying K times accrual factor α at each payment date. The term $P_t[T_0, T_n]$ is called the ***pv01*** of the swap, the present value of receiving 1 times α at each payment date. (Our notation for the value of the swap leg for simplicity suppresses information about the dates of the underlying swap.)

Since the floating leg is a series of regular libor payments, we can value it by replacing each libor with its forward. Thus the value of the floating leg is

$$V^{FL}(t) = \sum_{i=1}^{n} L_t[T_{i-1}, T_i]\alpha Z(t, T_i) = \sum_{i=1}^{n} (Z(t, T_{i-1}) - Z(t, T_i)) = Z(t, T_0) - Z(t, T_n),$$

the difference between two zero coupon bond prices, being the value of receiving 1 at the beginning of the swap and paying 1 at its end.

Therefore, we have the intuitively appealing result that the value of receiving libor payments on, say, one dollar is equal to the value of receiving one dollar at the beginning of the stream and giving it back at the end. We can simply take the dollar and repeatedly invest in a series of libor deposits.

The ***forward swap rate*** at time t for a swap from T_0 to T_n is defined to be the value $y_t[T_0, T_n]$ of the fixed rate K such that the value of the swap at t is zero.

Result *Equating fixed and floating legs we obtain for $t \leq T_0$*

$$y_t[T_0, T_n] = \frac{\sum_{i=1}^{n} L_t[T_{i-1}, T_i]\alpha Z(t, T_i)}{\sum_{i=1}^{n} \alpha Z(t, T_i)} = \frac{Z(t, T_0) - Z(t, T_n)}{P_t[T_0, T_n]}.$$

The forward swap rate is thus a weighted average of forward libors, which collapses to a ratio of a linear combination of ZCB prices to the pv01 (itself a linear combination of ZCB prices). Both formulations are useful later, in Chapters 12 and 15.

Result *The value $V_K^{SW}(t)$ at time $t \leq T_0$ of a swap where we pay a fixed rate K and receive libor is given by*

$$V_K^{SW}(t) = (y_t[T_0, T_n] - K)P_t[T_0, T_n].$$

Proof A complete proof is left to the exercises, using the fact that the value at t of the floating leg of the swap is by definition equal to the value of a fixed leg with fixed rate $y_t[T_0, T_n]$. ■

Compare this to the value of a simple forward contract

$$V_K(t, T) = (F(t, T) - K)e^{-r(T-t)}.$$

4.3 Spot-starting swaps

When $t = T_0$, $y_{T_0}[T_0, T_n]$ is called the ***par*** or ***spot-starting*** swap rate of maturity $T_n - T_0$. Given par swap rates for all T_i, we can recover $Z(T_0, T_i)$ for all i, a process known as bootstrapping. An example of this is given in the exercises. Swaps are now so liquid and widely traded that swap rates have become the 'primitive' source of interest rate price information from which ZCB prices, forward libor and other forward rates are typically calculated.

Result *The par swap rate $y_{T_0}[T_0, T_n]$ is the fixed rate at which one can invest 1 at time T_0 until time T_n, receiving fixed payments of $\alpha y_{T_0}[T_0, T_n]$ at times T_i, then notional 1 back at T_n.*

Proof At each T_{i-1} we can deposit 1 at $L_{T_{i-1}}[T_{i-1}, T_i]$ receiving $(1 + \alpha L_{T_{i-1}}[T_{i-1}, T_i])$ at T_i. We receive the payment of $\alpha L_{T_{i-1}}[T_{i-1}, T_i]$ and reinvest the notional 1 at T_i. We thus start with one unit of cash and obtain a payment stream of $\alpha L_{T_{i-1}}[T_{i-1}, T_i]$ at

T_i, for $i = 1, \ldots, n$, and 1 at T_n. By definition, this stream has the same value at T_0 as receiving a fixed payment stream of $\alpha y_{T_0}[T_0, T_n]$ at each date T_i, and 1 at T_n. ■

Hence the swap rate is a 'coupon' rate (where interest coupons received cannot necessarily be assumed to be reinvested at the same rate), distinct from the zero rate defined in Chapter 1, where we assume compounding at the same rate. Question 2 in the exercises explores this distinction.

4.4 Swaps as difference between bonds

Recall that a ***fixed rate bond*** with notional N and coupon c pays αcN at fixed dates T_i, and N at T_n, where we have $T_{i+1} = T_i + \alpha$. Let the price of the fixed rate bond at current time t be denoted $B_c^{FXD}(t)$.

A ***floating rate bond*** with notional N pays libor coupons $\alpha NL_{T_{i-1}}[T_{i-1}, T_i]$ at T_i, and N at T_n, and we denote its price $B^{FL}(t)$. We are suppressing information about the maturity of the bond in our price notation.

Consider a swap with notional 1 where we pay a fixed rate K and receive libor.

Result

$$V_K^{SW}(t) = V^{FL}(t) - V_K^{FXD}(t) = (V^{FL}(t) + Z(t, T_n)) - (V_K^{FXD}(t) + Z(t, T_n))$$
$$= B^{FL}(t) - B_K^{FXD}(t),$$

the difference between the price of a floating rate bond and the price of a fixed rate bond, as shown in Figure 4.2.

It follows that the value of a swap is bounded, and we prove this in Question 1(b). Note that $V^{FL}(t) + Z(t, T_n) = Z(t, T_0)$. Thus if $t = T_0$, then the floating rate bond has price 1, regardless of the level of interest rates. In other words, a floating rate bond has no

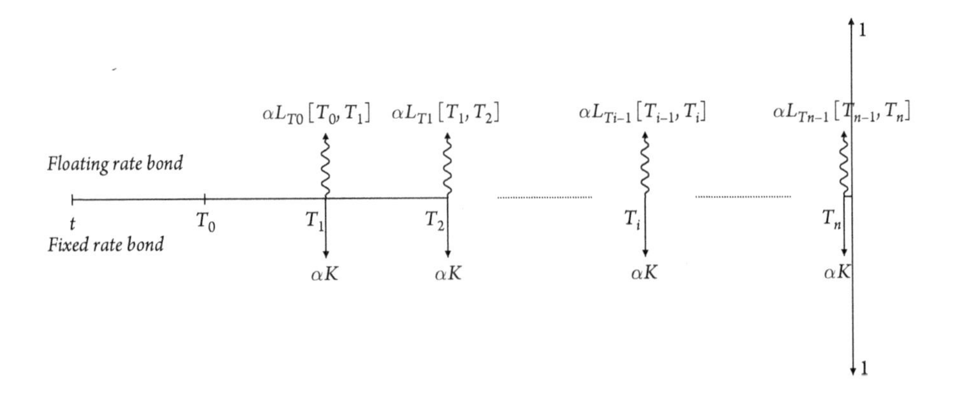

Figure 4.2 A swap as difference between two bonds

interest rate exposure. You should think carefully about the distinction between a fixed rate bond, whose cashflows are known but whose value changes, and a floating rate bond, whose cashflows are unknown but whose value (at coupon payment dates) is always 1.

Setting $t = T_0$ we have

$$V_K^{SW}(T_0) = (y_{T_0}[T_0, T_n] - K)P_{T_0}[T_0, T_n]$$

and also

$$V_K^{SW}(T_0) = 1 - B_K^{FXD}(T_0).$$

So we have

$$1 - B_K^{FXD}(T_0) = (y_{T_0}[T_0, T_n] - K)P_{T_0}[T_0, T_n].$$

Therefore, a bond whose coupon equals the swap rate of the same maturity has price 1 (or 'par'). A bond whose coupon is lower than the swap rate has price lower than par since if $y_{T_0}[T_0, T_n] > K$, then $B_K^{FXD}(T_0) < 1$. An exercise provides some more insight into fixed rate bonds.

4.5 EXERCISES

1. **Swap value**

   (a) Prove that the value $V_K^{SW}(t)$ at time $t \leq T_0$ of a swap from T_0 to T_n, where we pay fixed rate K and receive libor, is

   $$V_K^{SW}(t) = (y_t[T_0, T_n] - K)P_t[T_0, T_n].$$

   Hint Replace the floating leg of the swap with a fixed leg of equal value, or construct a portfolio of two swaps of the same maturity, but with different fixed rates.

   (b) By expressing a swap as a difference between a floating rate bond and fixed rate bond, prove that, for a given K, $V_K^{SW}(t)$ is bounded, that is, there exists finite l and u independent of interest rates such that $l \leq V_K^{SW}(t) \leq u$. For $t = T_0 = 0$ (a spot-starting swap), $T_n = n$ and frequency $\alpha = 1$, find bounds in terms of n and K.

   (c) Is the value of a forward contract on a stock S_t necessarily bounded above and below? Explain the key difference between the value of a swap and a stock forward contract.

2. **Bootstrapping and IRR discounting**

   Suppose the current one-year euro swap rate $y_0[0, 1]$ is 1.47%, and the two-year and three-year swap rates are 2.05% and 2.38% respectively. Euro swap rates are quoted with annual payments and 30/360 daycount (thus $\alpha = 1$).

(a) Use bootstrapping to calculate $Z(0,1)$, $Z(0,2)$ and $Z(0,3)$, and obtain $P_0[0,3]$, the present value of a three-year annuity paying €1 per year.

(b) Recall that the (annually compounded) zero rate for maturity T is the rate r such that $Z(0,T) = \frac{1}{(1+r)^T}$. Calculate the one-year, two-year and three-year zero rates, and compare them to the swap rates. In general, for upward sloping yield curves, that is, when $y_0[0,T_2] > y_0[0,T_1]$ for $T_2 > T_1$, will zero rates be higher or lower than swap rates?

(c) An approximate short-cut sometimes used on trading desks to calculate the present value of, say, a three-year annuity is to discount each payment at the three-year swap rate. In other words, it is assumed that $Z(0,n) = \frac{1}{(1+y_0[0,3])^n}$ for $n = 1, 2, 3$. This is often called ***IRR*** (internal rate of return) discounting. Calculate the error in valuing the annuity in (a) this way.

(d) Calculate the price of a three-year fixed rate bond of notional €1 and annual coupons of 2.38% using the ZCB prices calculated in (a), and verify this equals the price obtained via IRR discounting at a rate of 2.38%.

(e) The value of a swap with fixed rate K (see Question 1(a)) can be thought of as an annuity of amount $y_0[0,T_n] - K$ for period 0 to T_n. In yet another Wall Street quirk, euro swaps embedded in certain contracts are occasionally valued for cash settlement using IRR discounting at the current swap rate $y_0[0,T_n]$, rather than correct valuation using ZCBs. The logic for this originally was to reduce disagreements between banks on cash settlement of swaps. The swap rate is easily observed and IRR discounting is then a deterministic formula, whilst the bootstrapping undertaken in (a) was deemed too complicated. Suppose all euro swaps suddenly moved to this type of valuation. Given your answers to Question 2(a) to (c), comment briefly on whether you would expect euro swap rates to rise, fall or stay the same.

3. **Carry and rolldown**

Suppose euro swap rates are as given in Question 2. A hedge fund (HF) executes the following two trades with a dealer. (1) The HF pays fixed and receives floating on €100 million notional of a one-year swap. (2) The HF receives fixed and pays floating on €100 million notional of a three-year swap. Assume bid-offer costs are negligible.

(a) After one year, what net cashflow has the dealer paid to (or received from) the HF?

(b) Suppose after one year, one-year and two-year euro swap rates are unchanged. What is the current value of the remaining part of the HF trade?

(c) Suppose after one year, the one-year euro swap rate is unchanged but the two-year euro swap rate is now Y%. What value of Y gives a total zero profit on the trade (at $T = 1$)?

(d) Do you like the trades the HF executed? Discuss briefly the risks of the trade, in particular commenting on which interest rates the HF is exposed to.

4. **Swap frequency**
We have assumed that payment dates for the fixed and floating legs of a swap contract are the same. However, in practice, the frequency may differ—for example, in the US swap market the fixed leg is usually semi-annual ($\alpha = 0.5$), versus a floating leg of three-month libor paid quarterly ($\alpha = 0.25$).
   (a) Prove that the theoretical value of a swap is independent of the floating frequency—for example, three-month libor paid quarterly has the same value as six-month libor paid semi-annually.
   (b) Prove that the value of a swap with fixed rate K is not independent of the fixed leg frequency. In particular, does $P_t\left[T_0, T_n\right]$ increase or decrease if α changes from 0.25 to 1? Deduce directly whether the annual swap rate for T_0 to T_n is higher or lower than the quarterly swap rate for the same period. Assume the same daycount convention.

5. **Bonds and yield**
A fixed rate bond with notional 1 pays annual coupons of c at times T_i, for $i = 1, \ldots, n$, where $T_{i+1} = T_i + 1$, and notional 1 at time T_n.
   (a) Write down the bond price $B_c^{FXD}(t)$ at time $t \leq T_0$ in terms of ZCBs.
   (b) Suppose $t = T_0 = 0$. The ***yield*** of the bond is defined as the value Y such that

$$B_c^{FXD}(0) = \sum_{i=1}^{n} c \frac{1}{(1+Y)^i} + \frac{1}{(1+Y)^n},$$

   that is, the rate at which IRR discounting gives the bond price.
   By summing the geometric series, prove that

$$B_c^{FXD}(0) = 1 \Longleftrightarrow Y = c.$$

   (c) By writing a swap as the difference between a fixed rate bond and floating rate bond, prove that

$$B_c^{FXD}(0) = 1 \Longleftrightarrow c = y_0\left[0, T_n\right].$$

   Hint Recall the definition of the swap rate as the fixed rate that gives the swap zero value.

   Note We have, therefore, the interpretation of the T-year spot swap rate as the bond coupon such that a T-maturity bond has price par (that is, 100% of notional).

5

Futures contracts

A futures contract is a derivative with many similarities to a forward, essentially being a contract to trade an underlying asset at a fixed time in the future. However, there are two key differences. Most importantly, a futures contract involves cashflows each day up until the maturity date T and not just at T. Secondly, virtually all futures contracts are traded on exchanges rather than as bilateral over-the-counter contracts between two counterparties. Futures contracts on major government bonds and equity indices are very liquid, with hundreds of thousands of contracts trading each day.

5.1 Futures definition

Similarly to a forward contract, a ***futures contract*** (or ***future***) has a specified maturity T, an underlying asset, for example a stock with price at t denoted S_t, and a ***futures price*** $\Phi(t, T)$ at which one can go long or short the contract at no cost at time t. The futures price at maturity T is defined to be $\Phi(T, T) = S_T$.

The difference between a future and a forward lies in the cashflows during the life of a contract. In particular, the holder of a futures contract receives (or pays) changes in the futures price over the life of the contract, and not just at maturity. We below make precise the differences between the cashflows of the two contracts.

Forward contract. At time t, we can go long a forward contract with delivery price $F(t, T)$ at no cost. At maturity we receive (pay if negative) $S_T - F(t, T) = F(T, T) - F(t, T)$. There are no payments in between.

Futures contract. At time t, we can go long a futures contract with price $\Phi(t, T)$. Each day we receive (pay if negative) the ***mark-to-market*** change $\Phi(i, T) - \Phi(i - 1, T)$, where $\Phi(i, T)$ is the futures price on day i. This amount is also known as the ***variation margin***. Over the life of the contract we receive mark-to-market gains that total $\Phi(T, T) - \Phi(t, T) = S_T - \Phi(t, T)$. However, the constituent payments are made at different times and thus the value of the payments at T will not in general equal $S_T - \Phi(t, T)$.

Futures contracts are traded on electronic exchanges such as CME (formerly the Chicago Mercantile Exchange), CBOT (Chicago Board of Trade), NYMEX (New York Mercantile Exchange) and LIFFE (London International Financial Futures and Options Exchange, pronounced 'life'). There have been many recent mergers amongst exchanges. Futures have standardized monthly or quarterly maturity dates set by established market conventions. Contracts are netted, meaning that if we go long one contract then short one contract then we have no position.

Each market participant deposits initial margin at the exchange, and receives (or posts) additional variation margin as prices move up (or down). Initial margin is usually established to be a size that can cover 99% of five-day moves. It is important to note that margin—and in particular variation margin—accrues interest. This key feature gives rise to the difference between futures and forward contracts which we explore further in the next section.

5.2 Futures versus forward prices

Result *If interest rates are constant then*

$$\Phi(t, T) = F(t, T).$$

Note This is not a useful result for interest rate futures!

Proof Without loss of generality let $t = 0$, and assume that the variation margin is paid every Δ years, and that the maturity $T = n\Delta$. In practice, marking-to-market takes place every day so $\Delta = 1/365$. Let the constant continuously compounded interest rate be r, and consider the following trading strategy.

At time 0, go long $e^{-r(n-1)\Delta}$ futures contracts at futures price $\Phi(0, T)$.

At time Δ, increase position to $e^{-r(n-2)\Delta}$ contracts at futures price $\Phi(\Delta, T)$ (note that by definition of the futures price, we can do this at no cost).

At time $i\Delta$, for $i = 2, \ldots, n-2$, increase position to $e^{-r(n-i-1)\Delta}$ contracts at futures price $\Phi(i\Delta, T)$.

At time $(n-1)\Delta$, increase position to 1 contract.

With this strategy we receive the following amounts.

At time Δ we receive a mark-to-market gain of $(\Phi(\Delta, T) - \Phi(0, T))\, e^{-r(n-1)\Delta}$. This we can invest at rate r, and thus it compounds up by $e^{r(n-1)\Delta}$ to time $n\Delta$.

At time $(i+1)\Delta$ we receive a mark-to-market gain of $(\Phi((i+1)\Delta, T) - \Phi(i\Delta, T))e^{-r(n-i-1)\Delta}$, which compounds up by $e^{r(n-i-1)\Delta}$ to time $n\Delta$.

Therefore, the value at time $T = n\Delta$ of the cashflows received from mark-to-market gains or losses is

$$\begin{aligned}&\sum_{i=0}^{n-1} (\Phi((i+1)\Delta, T) - \Phi(i\Delta, T))\, e^{-r(n-i-1)\Delta} e^{r(n-i-1)\Delta} \\ &\quad = \Phi(n\Delta, T) - \Phi(0, T) = S_T - \Phi(0, T).\end{aligned}$$

So an initial holding of $e^{-rT}\Phi(0,T)$ of cash, plus the costless futures trading strategy given above, results in a portfolio with value at T

$$\Phi(0,T) + (S_T - \Phi(0,T)) = S_T.$$

However, consider the portfolio consisting of $e^{-rT}F(0,T)$ cash invested at r, plus one forward contract. This is also worth S_T at time T. Therefore,

$$\Phi(0,T) = F(0,T). \qquad \blacksquare$$

Consider now the case where the interest rate r is not constant. Suppose that the asset price S_T is positively correlated with the interest rate and thus tends to increase as interest rates increase. If we are long a futures contract, then we receive mark-to-market gains earlier than the forward in environments when interest rates are high, and thus the gains can be invested at a higher rate. Similarly, losses from the long position have to be paid early when rates are low (and thus the future value of the losses is lower). Thus in this case, we would prefer to hold a long futures position relative to a long forward position. Hence, if the asset price and interest rate are positively correlated, we deduce by this heuristic argument that $\Phi(t,T) - F(t,T) > 0$. The difference $\Phi(t,T) - F(t,T)$ is called the ***futures convexity correction***.

A more detailed analysis of the futures convexity correction is given in Hunt and Kennedy (2004). When the interest rate is random, then the money market account is also random. We state here the following general result, which makes precise our earlier heuristic argument. Its proof is beyond our scope.

Result *$\Phi(t,T) - F(t,T)$ is proportional to the covariance between the asset price S_T and money market account M_T. In particular, if the asset price S_T and money market account M_T are independent, then $\Phi(t,T) = F(t,T)$. If they are positively correlated, then $\Phi(t,T) - F(t,T) > 0$.*

The covariance between the asset price and the money market account is proportional to $(T - t)$, so typically the convexity correction is greater for futures contracts with longer maturities. The convexity correction tends to zero as t approaches T.

The most liquid futures contracts in the world are futures on US treasuries (the ***bond future***) and German government bonds (the ***bund future***). Here, the bond price and interest rate are almost always negatively correlated. (Can you think of a situation when this may not hold?) Therefore, the bond price and money market account are negatively correlated, and so in general $\Phi(t,T) - F(t,T) < 0$. However, since most liquid bond futures have maturities of three months or less, the size of this convexity correction is often small.

5.3 Futures on libor rates

One common type of futures contract are those linked to libor rates, in particular ***Eurodollar*** futures on three-month US dollar libor, ***short sterling*** futures on three-month sterling libor and ***Euribor*** futures on three-month Euribor.

Note In a particularly confusing case of market terminology, Eurodollar futures have nothing to do with the euro dollar FX rate.

Note To create more confusion, Euribor (Euro Interbank Offered Rate) is a subtly different interest rate to euro libor. The mechanism for setting euro libor is similar to that for US dollar libor or sterling libor: the rate is set at 11am in London, based on a poll of rates from 15 international contributor banks active in the London interbank market. Euribor is a complementary rate established by the European Banking Federation, set at 11am in Brussels, one hour earlier than libor. A broader panel of 43 mostly European banks is used. Small differences may occur between Euribor and euro libor, due to the different credit quality of the banks in the two panels. Most European interest rate derivatives—and in particular Euribor futures—reference Euribor not euro libor.

The ***Eurodollar futures*** contract with settlement date T is defined by its price at maturity

$$\Phi(T, T) = 100(1 - L_T[T, T + \alpha]),$$

where $L_T[T, T + \alpha]$ is the three-month dollar libor rate. For example, if three-month dollar libor on the maturity date of the futures contract fixes at 1.25%, then the futures price at maturity is 98.75.

Eurodollar futures trade with quarterly maturities out to ten years, maturing on IMM dates (see Chapter 1), the third Wednesdays of March, June, September and December, plus monthly maturities for the first four intermediate months. One Eurodollar contract is defined to have a 'tick size' of \$25. That is, if we are long one future and its price moves from 98.75 to 98.76 (equivalent to an interest rate moving by one ***basis point***, 0.01%), then the mark-to-market gain is \$25. We can thus consider the effective notional of one Eurodollar futures contract to be \$1 million, since one basis point of \$1 million multiplied by a daycount fraction of one quarter equals \$25. The notional size varies for different futures contracts, for example, the effective notional of a short sterling future is £500,000. Having precise knowledge of the exact contract specification for each futures contract is essential, and many trading mishaps have resulted from incomplete knowledge of the characteristics of a contract.

The holder of a long position in a Eurodollar futures contract receives positive variation margin when forward interest rates go down, and the futures price goes up. Forward interest rates and the money market account are usually positively correlated. Therefore, a long Eurodollar position receives money early when interest rates are low, and pays out money early when rates are high. Thus, the Eurodollar futures price will typically be lower than the forward price. Specifically, we have

$$\Phi(t, T) < 100(1 - L_t[T, T + \alpha]),$$

and thus

$$\left(1 - \frac{\Phi(t, T)}{100}\right) - L_t[T, T + \alpha] > 0.$$

This quantity is known as the ***Eurodollar convexity correction***.

Note As of July 2012, the December 2015 futures contract (then currently the fourth outstanding December contract) has price 98.925. The forward libor for that date is 1.04%, so the convexity correction is 3.5bp. Which is the better predictor of where libor will be in December 2015?

Note Eurodollar and other futures contracts on short-dated interest rates are often referred to by a colour code. The first four quarterly contracts are called the whites (or fronts), followed by the reds (the second four quarterly contracts), greens, blues, golds, purples, oranges, pinks, silvers and coppers. Contracts beyond the golds are typically less liquid and thus the colour coding beyond the golds is used infrequently. As of July 2012, the December 2015 contract would be referred to as 'blue December'.

Note Early in my finance career, I tried to persuade my boss that the Eurodollar convexity correction should be zero. I had a proof that the forward price equals the futures price, I claimed, being eager to show off my mathematical acumen. It was humbling to have it pointed out that the proof depended on interest rates being constant; hardly a valid assumption for Eurodollar futures, which move continually.

Since the convexity correction is proportional to the covariance of $1 - L_T[T, T + \alpha]$ and the money market account, it tends to be larger for long-dated Eurodollar futures such as the blues and golds, than for the reds and greens. In addition, given we know that the determination of forward libor does not depend on the distribution of $L_T[T, T + \alpha]$, we can conclude that the price of the Eurodollar future does indeed depend on this distribution. In particular, futures contracts have non-zero 'vega' exposure, meaning that their price is a function of the standard deviation of the underlying rate. We explore the concept of vega further in Chapter 10.

5.4 EXERCISES

1. **Eurodollar futures versus FRAs**
   (a) Suppose in September 2011 we observe that the Eurodollar futures prices for the March 2012, June 2012 and September 2012 contracts are all 98.00. Consider FRAs with these three maturities. Which FRA is likely to have the lowest rate?
   (b) Suppose in September 2011, the March 2013 Eurodollar contract is 97.25 and the FRA for this date is quoted at 2.65%. If you are told that the futures price will not move over its life, what trade would you do? How about if you were told the FRA rate would not move?

2. Futures on bonds and stocks

(a) Is the futures price of a fixed rate bond likely to be higher, lower or the same as its forward price? Give brief reasoning.

(b) Is the futures price (for a futures contract with maturity $T < T_0$) of a floating rate bond likely to be higher, lower or the same as its forward price? How does your answer change if $T = T_0$?

(c) Is the futures price of a stock likely to be higher, lower or the same as its forward price?

6

No-arbitrage principle

In this chapter we formalize the no-arbitrage and replication arguments used when determining forward prices. These arguments have intuitive appeal, and become a powerful component of the pricing machinery for more complicated derivative contracts such as options. Whilst the definition of arbitrage and arbitrage portfolios in continuous time (in particular when infinitely many trades can be made) is non-trivial, the key ideas behind no-arbitrage can be well established in the discrete setting, and we proceed accordingly here. As mentioned, finance is in practice executed in discrete time and value.

6.1 Assumption of no-arbitrage

Let the value at current time t of a portfolio **A** of assets be denoted $V^A(t)$.

For some future time $T > t$, the value of the portfolio $V^A(T)$ is a random variable, and we can write its value as $V^A(T, \omega_i)$ for the sample outcome ω_i. Here, we assume there is a discrete sample space of possible outcomes at T, denoted by $\Omega = \{\omega_i\}$ for $i \in I$, a discrete index set.

A portfolio **A** is an ***arbitrage portfolio*** if its current value $V^A(t) \leq 0$ and, for some $T > t$, $V^A(T) \geq 0$ for all states of the world, with $V^A(T) > 0$ with non-zero probability. That is, an arbitrage portfolio satisfies $V^A(T, \omega_i) \geq 0$ for all i, and $P\{\omega_j : V^A(T, \omega_j) > 0\} > 0$ for some j.

The ***assumption of no-arbitrage***—that is, there exist no arbitrage portfolios—underpins quantitative finance. This assumption is often informally restated as the phrases 'there is no such thing as free money' or 'one cannot get something for nothing'.

Example The portfolio **A** consisting of short one forward contract with maturity T at the forward price $F(t, T)$, a loan of S_t and long one stock is an arbitrage portfolio if $F(t, T) > S_t e^{r(T-t)}$, since $V^A(t) = 0$ and $V^A(T, \omega_i) = F(t, T) - S_t e^{r(T-t)} > 0$ for all i. The assumption of no-arbitrage thus implies that $F(t, T) \leq S_t e^{r(T-t)}$. A similar argument gives $F(t,T) \geq S_t e^{r(T-t)}$. Therefore, the no-arbitrage assumption allows pricing of forwards with no assumptions about the distribution of the stock price S_T.

6.2 Monotonicity theorem

Monotonicity theorem. *Assume no-arbitrage. If portfolios* **A** *and* **B** *are such that* $V^A(T,\omega_i) \geq V^B(T,\omega_i)$ *for all i, then* $V^A(t) \geq V^B(t)$. *If in addition* $V^A(T,\omega_j) > V^B(T,\omega_j)$ *for some j with* $P\{\omega_j\} > 0$, *then* $V^A(t) > V^B(t)$.

Proof Apply the no-arbitrage assumption to the portfolio **C** = **A** – **B**. This assumes we can go short—hold negative amounts of assets—at will. We have

$$V^C(T,\omega_i) \geq 0 \text{ for all } i.$$

If $V^C(t) = -\epsilon < 0$, then a portfolio of **C** plus ϵ of cash is an arbitrage portfolio, since it has zero value at time t and necessarily positive value $\geq \epsilon$ at time T. So we conclude that

$$V^C(t) \geq 0 \Rightarrow V^A(t) \geq V^B(t).$$

In addition, if $V^C(T,\omega_j) > 0$ for some j with $P\{\omega_j\} > 0$, then $V^C(t) > 0$, else **C** itself is an arbitrage portfolio. ■

Example A derivative that has non-negative payout at T has non-negative value at all times $t \leq T$. We have provided a formal argument for an almost tautological result.

We now can immediately prove an important corollary.

Corollary to monotonicity theorem. *If* $V^A(T,\omega_i) = V^B(T,\omega_i)$ *for all i, then* $V^A(t) = V^B(t)$.

This corollary is the formal statement of the replication argument from Chapter 2. We refer to the monotonicity theorem and its corollary repeatedly later in the book.

Example If Z_1 and Z_2 are ZCBs with the same maturity T, then $Z_1(t,T) = Z_2(t,T)$ for all t. The proof is obvious since $Z_1(T,T) = Z_2(T,T) = 1$.

Example Let us revisit the FRA, where we considered two portfolios. Portfolio **A** consisted of one ZCB maturing at T and $-(1+\alpha K)$ ZCBs maturing at $T+\alpha$, plus a libor deposit at time T at rate $L_T[T, T+\alpha]$. Portfolio **B** consisted of long one FRA with a fixed rate K. Hence

$$V^A(T+\alpha) = V^B(T+\alpha) = \alpha(L_T[T,T+\alpha] - K).$$

Therefore,

$$V^A(t) = V^B(t) \text{ for all } t \leq T+\alpha.$$

Note that portfolio **A** had a dynamic element, namely investing the one unit of cash from the ZCB maturing at T in a libor deposit at time T. This strategy can be carried out at zero cost, since the deposit is at the current market libor rate, without any asset of non-zero value being added to or subtracted from the portfolio. Such a strategy is termed ***self-financing***. The trading strategy that showed the futures price equalled the forward price was also self-financing, since it involved trading futures at the then current futures price. Going long a forward at its forward price, paying fixed on a swap at the forward swap rate or borrowing money at the current interest rate are all self-financing trades. We formally need to include the self-financing condition in our definition of an arbitrage portfolio. Otherwise, for example, the portfolio which is empty at time 0 but to which cash $\epsilon > 0$ is added (without any liability) would trivially be an arbitrage portfolio. It is important to understand the differences between an empty portfolio which borrows cash ϵ at the current interest rate r for time T, and an empty portfolio to which ϵ cash is added. The latter is not self-financing.

6.3 Arbitrage violations

The assumption of no-arbitrage is a fundamental foundation of quantitative finance. Assuming no-arbitrage, we are able, for example, to order derivative prices via the monotonicity theorem based on knowledge of their payouts at maturity. The edifice of quantitative finance was constructed upon this assumption and we will adopt it throughout the remainder of the book as we build our theory and results.

However, financial markets in practice have a tendency to challenge foundational assumptions, and key challenges for the field have often arisen from occasions when no-arbitrage arguments were violated. As mentioned in Chapter 2, the period of the financial crisis following the Lehman Brothers bankruptcy in particular contained several examples of extreme price movements and violations of arbitrage bounds that stunned seasoned practitioners. We examine here one stark and relatively straightforward instance from 2008–2009. A further example is detailed in Chapter 13.

Example Consider again the example of two bonds with the same maturity date. Figure 6.1 shows the difference in yield, measured in basis points, between two US government bonds, both of which have maturity date 15 August 2015. We can see these bonds generally traded with a very small yield differential until late 2008. During the financial crisis, however, their yields diverged dramatically and remained apart for several months, essentially violating our result $Z_1(t, T) = Z_2(t, T)$. The yield differential normalized following government policy responses in early 2009. Further details of this episode and possible explanations for the observed prices are given in Taliaferro and Blyth (2011). Note that the data are yields of coupon bonds, but these can easily be restated in terms of ZCB prices.

Note I like to draw an analogy here with pure mathematics. Underpinning set theory is an assumption about choosing elements from an uncountably infinite number of sets called the ***axiom of choice***. Given this axiom, it is possible to establish a coherent theory which allows, for example, the 'well-ordering' of the real numbers; that is,

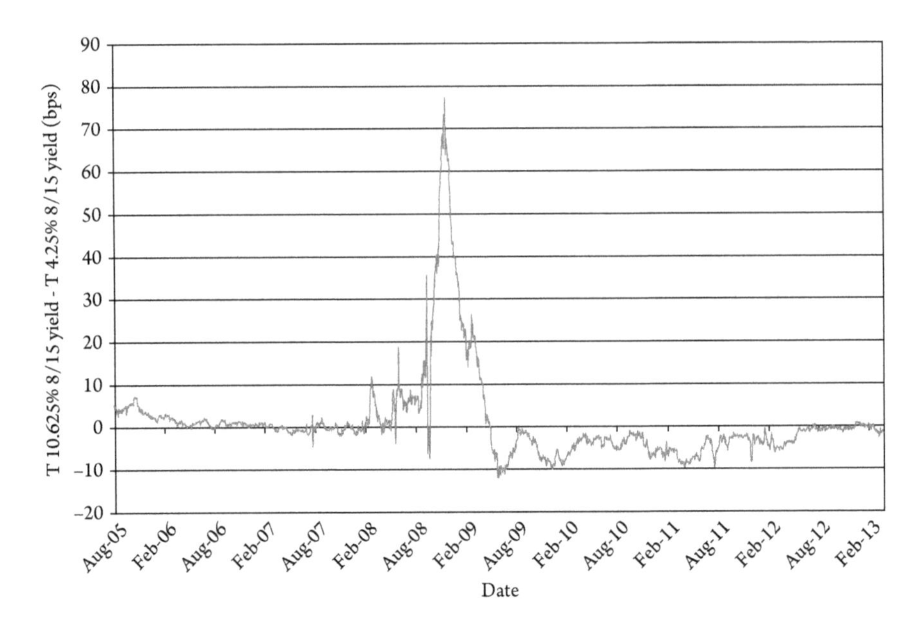

Figure 6.1 Yield difference between two bonds with August 2015 maturity

that any non-empty subset of the real numbers has a least element under a particular ordering. Without the axiom of choice, set theory becomes far more complex and messy. In finance, the assumption of no-arbitrage plays an analogous role, allowing the establishment of bounds on and ordering of derivative prices. Without the assumption, quantitative finance becomes far messier, and it becomes hard to order derivative prices. The financial crisis of 2008–2009 gave a flavour of what finance could look like without its own 'axiom of choice'. Further discussion is given in Blyth (2011).

6.4 EXERCISES

1. **Arbitrage portfolios**

   Which of the following necessarily imply a violation of the no-arbitrage assumption? Assume $T > 0$ and $\epsilon > 0$, and that all portfolios are self-financing.

   (a) A portfolio which has value ϵ today, and value 2ϵ at time T.

   (b) A portfolio which has zero value today, and expected positive value at T.

   (c) A portfolio which has value $-\epsilon$ today, and zero value at T.

   (d) A portfolio which has value $-\epsilon$ today, and expected positive value at T.

   (e) A portfolio which has zero value today, and value ϵ at T.

(f) A portfolio which has zero value today, and positive value at T for some sample outcomes with positive probability.

(g) A portfolio which has zero value today, always non-negative value at T, and positive value at T for some sample outcomes.

(h) A portfolio which has zero value today, always non-negative value at T, and expected positive value at T.

7

Options

We now introduce options, our second key class of derivative, which provide the building blocks for a broad range of derivative products.

7.1 Option definitions

A ***European call option*** with ***strike*** or ***exercise price*** K and ***maturity*** or ***exercise date*** T on an asset is the right—but not the obligation—to buy the asset for K at time T. The execution of this right is termed the ***exercise*** of the option.

Since we would only opt to pay K for an asset worth S_T if $S_T \geq K$, the payout $g(S_T)$ of a call option at time T is

$$\left.\begin{array}{ll} S_T - K & \text{if } S_T \geq K \\ 0 & \text{if } S_T \leq K \end{array}\right\} = \max\{S_T - K, 0\} = (S_T - K)^+.$$

The payout of a call option with strike 100 is shown in Figure 7.1.

A ***European put option*** is the right to sell the asset for K at time T. The payout of a put option at time T is

$$\left.\begin{array}{ll} K - S_T & \text{if } S_T \leq K \\ 0 & \text{if } S_T \geq K \end{array}\right\} = \max\{K - S_T, 0\} = (K - S_T)^+.$$

Figure 7.2 shows the payout of a put option with strike 90.

A ***straddle*** is a call plus a put of the same strike and maturity, and has payout $|S_T - K|$.

From simple beginnings, a world opens up.

A ***European*** option allows exercise only at T.

An ***American*** option allows exercise at any time $t \leq T$.

A ***Bermudan*** option allows exercise at a finite set of times $T_0, \ldots, T_n \leq T$.

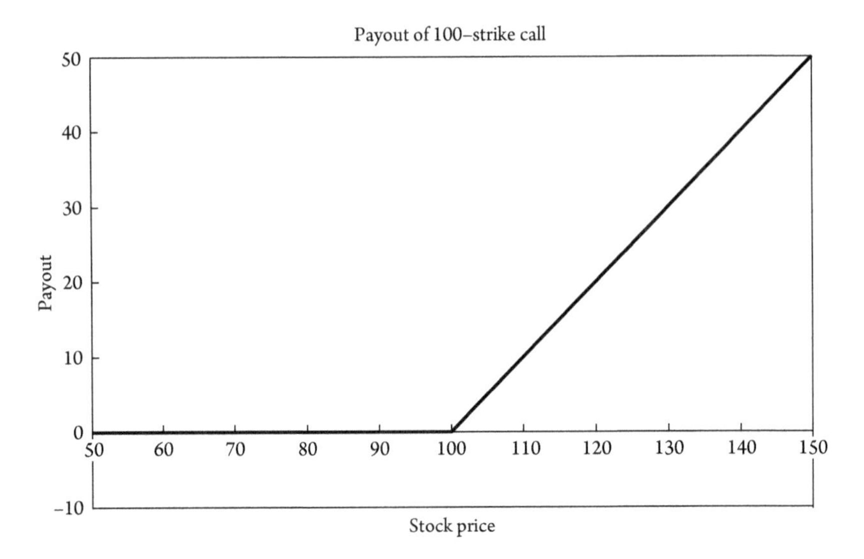

Figure 7.1 Payout of a call option

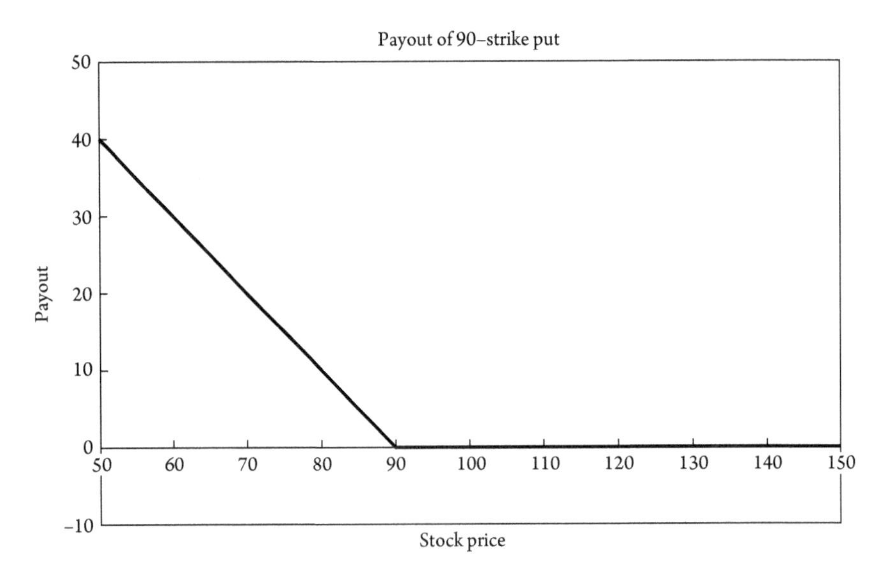

Figure 7.2 Payout of a put option

European options are common on FRAs, and known as caps and floors (Chapter 12). American options are common on stocks, and we investigate these in this chapter. Bermudan options are common on swaps, where they underlie mortgages and callable bonds (Chapter 13).

At time t, a call option with strike K at maturity T is said to be:

at-the-money if $S_t = K$;

in-the-money if $S_t > K$, that is, if the call option would have positive value if the stock price remains unchanged until maturity;

out-of-the-money if $S_t < K$, that is, if the option would be worthless if the stock price remains unchanged.

Note The phrases 'call with strike K', 'call struck at K', 'K-strike call' and 'K call' are used interchangeably.

At time $t \leq T$, a call option is struck ***at-the-money-forward (ATMF)*** if $K = F(t, T)$, in-the-money-forward if $K < F(t, T)$ and out-of-the-money-forward if $K > F(t, T)$. Note that an option which is ATMF at time t may not be ATMF at subsequent times.

The ***intrinsic value*** of an option is its payout were we able to exercise it now. Thus the intrinsic value of a call is $\max\{S_t - K, 0\}$, and of a put is $\max\{K - S_t, 0\}$. An option is in-the-money if it has positive intrinsic value.

Let us attempt a naïve initial valuation for a call option. Suppose $S_0 = 50$, and consider a one-year call option struck at 60. Suppose there are three possible states of the world at time $T = 1$, denoted $\omega_1, \omega_2, \omega_3$, such that $S_1(\omega_i) = 110, 70, 40$ for $i = 1, 2, 3$ respectively. The call option payout is 50, 10, and 0 in the three states respectively, as shown in Figure 7.3.

Suppose each outcome has probability $1/3$ of occurring, so the expected option payout $E(S_1 - K)^+ = 20$. Suppose further that $Z(0, 1) = 0.95$. The discounted present value of the expected payout is $Z(0, T)E(S_T - K)^+ = 19$. Is this a plausible candidate for the option price?

If the option price were indeed 19, we could sell one option, borrow an additional 31 of cash and buy one stock. This portfolio consisting of long one stock, short one call and a loan of 31 has zero value at time $T = 0$.

At $T = 1$ we must repay $31/Z(0, 1) = 32.63$ on the loan. The portfolio has the values $110 - 50 - 32.63 > 0$, $70 - 10 - 32.63 > 0$ and $40 - 32.63 > 0$ in states ω_1, ω_2, ω_3 respectively. Therefore, it is an arbitrage portfolio, and we conclude that an ad-hoc model for option pricing can rapidly lead to arbitrage.

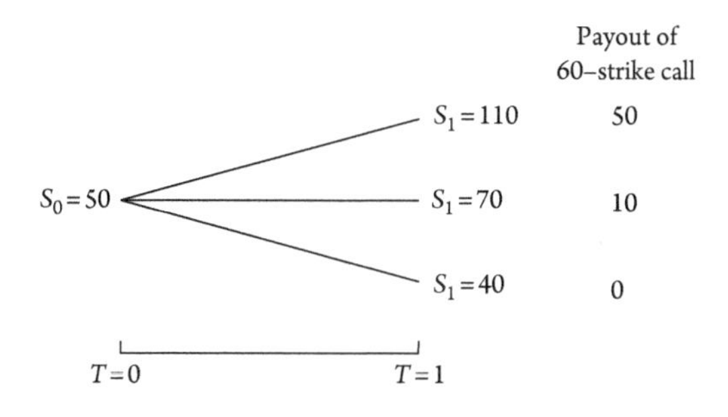

Figure 7.3 Call option payout

Before we attempt further to develop a framework for arbitrage-free pricing of options, we investigate a range of option properties.

7.2 Put-call parity

Let $C_K(t, T)$ be the price at time t of the European call, and $P_K(t, T)$ be the price of the European put, with strike price K and maturity T.

Result

$$C_K(t, T) \geq 0 \text{ and } P_K(t, T) \geq 0.$$

Proof The result follows immediately from the monotonicity theorem using the facts that

$$C_K(T, T) = (S_T - K)^+ \geq 0 \text{ and } P_K(T, T) = (K - S_T)^+ \geq 0. \qquad \blacksquare$$

Recall from Chapter 2 that $F(t, T)$ is the forward price for a forward contract with maturity T, $V_K(t, T)$ is the value at time t of the forward contract with delivery price K, and we have the relationship

$$V_K(t, T) = (F(t, T) - K)Z(t, T).$$

We now prove ***put-call parity***, which relates the prices of the call and put of the same strike with the value of the forward.

Result (put-call parity).

$$C_K(t, T) - P_K(t, T) = V_K(t, T).$$

Proof Consider a portfolio of long one call and short one put, both with strike K and maturity T. At time T, the payout of this portfolio (see Figure 7.4) is

$$\begin{cases} S_T - K - 0 & \text{if } S_T \geq K \\ 0 - (K - S_T) & \text{if } S_T \leq K. \end{cases}$$

So the payout is $S_T - K$, the payout of a forward contract. Therefore, by the corollary to the monotonicity theorem, the value of this portfolio at $t \leq T$ equals the value of the forward contract at t. ∎

Put-call parity states that long one call and short one put equals long one forward contract. Similarly, long one call equals long one forward and long one put. We can always convert from a call to a put by trading the forward, and we saw in Chapter 2 that the forward itself can be replicated by a holding of stock and cash.

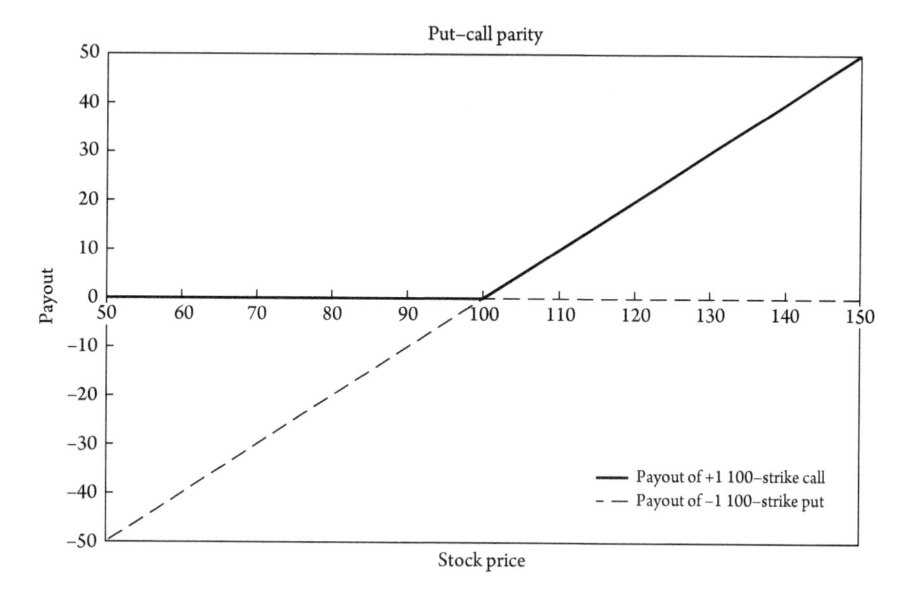

Figure 7.4 Put-call parity

For ATMF options put-call parity becomes

$$C_{F(t,T)}(t,T) - P_{F(t,T)}(t,T) = 0.$$

That is, the price of an ATMF put equals that of an ATMF call. This result is independent of any distributional assumptions and holds regardless of whether S_T is skewed to the left or right, or has any other unusual distributional features.

7.3 Bounds on call prices

Result *The European call price on a non-dividend paying stock satisfies*

$$\max\{0, S_t - KZ(t,T)\} \leq C_K(t,T) \leq S_t.$$

Note We show in Chapter 10 that both upper and lower bounds are tight, that is, can be attained.

Proof The upper bound follows immediately from the monotonicity theorem since $C_K(T,T) = \max\{S_T - K, 0\} \leq S_T$.

To establish the lower bound let portfolio **A** consist of long one call and K ZCBs with maturity T, and let portfolio **B** consist of long one stock.

At T, if $S_T \geq K$, the call is exercised (with value $S_T - K$) and portfolio **A** has value S_T. If $S_T < K$, the call expires worthless and portfolio **A** has value K at T.

Thus portfolio **A** has value at T equal to $\max\{S_T, K\} \geq S_T$, which is the value of portfolio **B** at time T. Therefore, by the monotonicity theorem,

$$C_K(t,T) + KZ(t,T) \geq S_t. \qquad \blacksquare$$

Note that since $Z(t,T) \leq 1$, $C_K(t,T) \geq S_t - K$, the intrinsic value of the option, and we have the following interesting corollary.

Corollary *The price of an American call and European call on a non-dividend-paying stock are equal.*

Proof Let $\widetilde{C}_K(t,T)$ be the price of the American call.
$\widetilde{C}_K(t,T) \geq C_K(t,T)$ is obvious.
To prove that $\widetilde{C}_K(t,T) \leq C_K(t,T)$, consider two cases.

If the American call is not exercised before T, then clearly $\widetilde{C}_K(t,T) = C_K(t,T)$.

If the American call is exercised early, say at time $t < T$, then

$$\widetilde{C}_K(t,T) = S_t - K \leq C_K(t,T). \qquad \blacksquare$$

The intuition behind this somewhat surprising result is that, regardless of when we exercise the American call, we end up with S_T at T. Therefore, we would rather pay the strike price K at later T than at some time $t < T$ (which is equal to paying $K/Z(t,T) > K$ at T).

The equality between American and European options does not hold for calls on a dividend paying stock. Here, by exercising early, we may receive dividends from the stock which we would not have received otherwise. American puts on stocks are also usually worth more than European puts, since by exercising early we receive the strike price K early and can invest this amount. In the exercises we establish bounds on put prices.

7.4 Call and put spreads

A ***call spread*** is a portfolio consisting of long one call option with strike K_1 and short one call option with strike K_2, both with maturity T, where $K_1 < K_2$.

A call spread has value $C_{K_1}(t,T) - C_{K_2}(t,T)$ at time t. At time T its payout equals

$$\begin{cases} 0 & \text{if } S_T \leq K_1 \\ S_T - K_1 & \text{if } K_1 \leq S_T \leq K_2 \\ K_2 - K_1 & \text{if } S_T \geq K_2, \end{cases}$$

as illustrated in Figure 7.5.

Result *If $K_1 < K_2$ then*

$$C_{K_1}(t,T) \geq C_{K_2}(t,T) \text{ and } P_{K_1}(t,T) \leq P_{K_2}(t,T).$$

In particular, the call spread and put spread have non-negative value for all $t \leq T$.

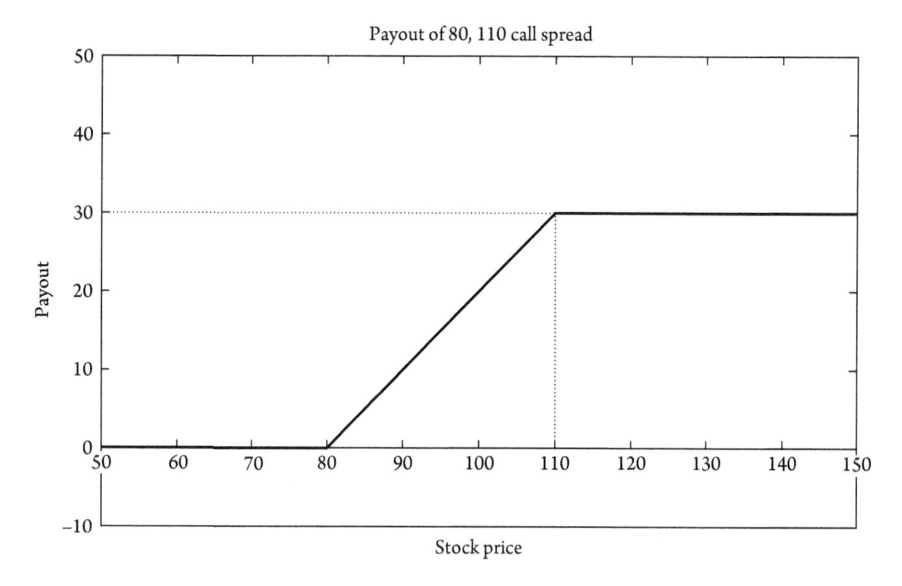

Figure 7.5 Call spread

Proof The result follows immediately from the monotonicity theorem, since the call spread has non-negative payout at maturity T. Alternatively, if $C_{K_1}(t,T) < C_{K_2}(t,T)$, then we can construct an arbitrage portfolio by selling one call with strike K_2 and buying one call with strike K_1. ∎

Result *If $K_1 < K_2$ then*

$$C_{K_1}(t,T) - C_{K_2}(t,T) \leq Z(t,T)(K_2 - K_1)$$

and

$$P_{K_2}(t,T) - P_{K_1}(t,T) \leq Z(t,T)(K_2 - K_1).$$

Proof By put-call parity

$$C_{K_1}(t,T) - P_{K_1}(t,T) = (F(t,T) - K_1)Z(t,T)$$

and

$$C_{K_2}(t,T) - P_{K_2}(t,T) = (F(t,T) - K_2)Z(t,T).$$

Therefore,

$$(C_{K_1}(t,T) - C_{K_2}(t,T)) + (P_{K_2}(t,T) - P_{K_1}(t,T)) = Z(t,T)(K_2 - K_1).$$

The first two terms are both non-negative. ∎

Result *Combining the two previous results we have*

$$C_{K_2}(t,T) \le C_{K_1}(t,T) \le C_{K_2}(t,T) + Z(t,T)(K_2 - K_1).$$

Note Since $Z(t,T)$ does not depend on K_1 or K_2, this result is equivalent to the mathematical statement that $C_K(t,T)$ is a Lipschitz continuous function of K with Lipschitz constant $Z(t,T)$.

Alternative Proof A portfolio consisting of one call with strike K_2 and $(K_2 - K_1)$ ZCBs with maturity T will have payout at T

$$(K_2 - K_1) + \max\{S_T - K_2, 0\} \ge \max\{S_T - K_1, 0\},$$

the payout of a portfolio consisting of one call with strike K_1. Therefore, the result follows by the monotonicity theorem. ■

We revisit call spreads in Chapter 11.

7.5 Butterflies and convexity of option prices

Result *Let $K_1 < K_2$, $\lambda \in (0,1)$, and let $K^* = \lambda K_1 + (1-\lambda)K_2$. Then we have*

$$C_{K^*}(t,T) \le \lambda C_{K_1}(t,T) + (1-\lambda)C_{K_2}(t,T).$$

In other words, $C_K(t,T)$ is a convex function of K.

Proof For $\lambda \in (0,1)$, consider a portfolio consisting of λ calls with strike K_1, $(1-\lambda)$ calls with strike K_2, and -1 call with strike K^*. At maturity T its value is

$$\begin{cases} 0 & \text{if } S_T \le K_1 \\ \lambda(S_T - K_1) & \text{if } K_1 \le S_T \le K^* \\ \lambda(S_T - K_1) - (S_T - \lambda K_1 - (1-\lambda)K_2) = (1-\lambda)(K_2 - S_T) & \text{if } K^* \le S_T \le K_2 \\ \lambda(S_T - K_1) - (S_T - \lambda K_1 - (1-\lambda)K_2) + (1-\lambda)(S_T - K_2) = 0 & \text{if } S_T \ge K_2 \end{cases}$$

which is always non-negative. Hence, by the monotonicity theorem, its value at $t \le T$ is non-negative, that is

$$\lambda C_{K_1}(t,T) + (1-\lambda)C_{K_2}(t,T) - C_{K^*}(t,T) \ge 0.$$ ■

A portfolio of this form (or multiples of it) is called a ***call butterfly***. We revisit them in Chapter 11, and see that they are closely related to probability density functions. Here we have shown by inspecting the butterfly payout at maturity, and a simple appeal to the monotonicity theorem, that $C_K(t,T)$ is convex. In practice, the call butterflies most commonly

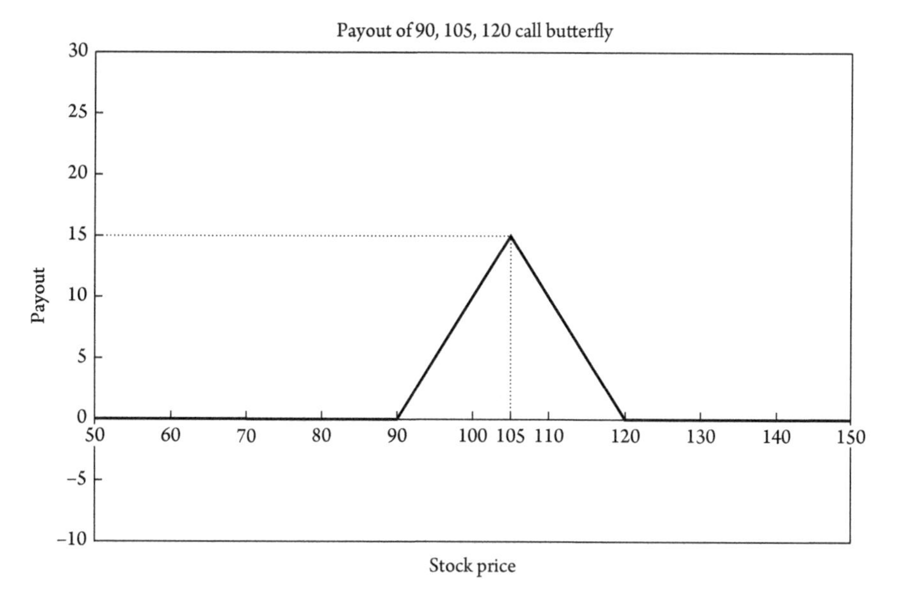

Figure 7.6 Call butterfly

traded are usually of the form: +1 call with strike K_1; −2 calls with strike $(K_1 + K_2)/2$; and +1 call with strike K_2. Figure 7.6 gives a typical example.

7.6 Digital options

A ***digital call option*** with strike K and maturity T has payout at T

$$\begin{cases} 1 & \text{if } S_T \geq K \\ 0 & \text{if } S_T < K. \end{cases}$$

The payout of a digital call is shown in Figure 7.7.

A ***digital put option*** with strike K and maturity T pays

$$\begin{cases} 1 & \text{if } S_T \leq K \\ 0 & \text{if } S_T > K. \end{cases}$$

Note The terms of the digital contract usually specify precisely whether or not the payout is made when the asset price at maturity exactly equals the strike. In our definitions above, we have specified that the payment is indeed made. For continuous distributions, the likelihood of this happening is infinitesimally small, although in certain markets where there are elements of discreteness (for example, libor rates are often quoted to the nearest eighth of a basis point), the price between the two versions can differ meaningfully.

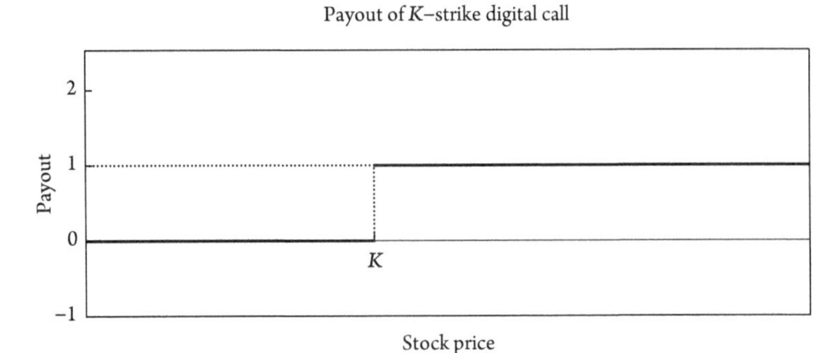

Figure 7.7 A digital call

7.7 Options on forward contracts

We now introduce options on forward contracts. The simplest versions are straightforward, yet subtleties soon arise. Their analysis provides a good exercise in understanding the precise terms of derivative contracts. Options on forwards are particularly prevalent in interest rate derivatives, where the underlying variable is a FRA.

Recall that a forward contract is an agreement to buy an asset at time T for fixed price K. Let $C^{F_K}(t, T)$ be the price of a European call option with exercise date T on a forward contract, specifically, the right to enter at time T (at zero cost) into a long forward contract with delivery price K and maturity T. Note that the exercise date of the option and the maturity date of the forward are the same.

At time T, the forward contract is simply a purchase of the asset. Thus the payout of the option at T is $(S_T - K)^+$, equal to the payout of the European call on the stock. Therefore,

$$C^{F_K}(t, T) = C_K(t, T).$$

So the right to enter (at zero cost) into a forward contract with delivery price K equals the right to buy a stock for K. Similarly,

$$P^{F_K}(t, T) = P_K(t, T),$$

where P^{F_K} is the price of the option to enter at time T into a short forward contract with delivery price K and maturity T.

Whilst European options on forwards are equal in value to European options on the underlying asset, American options on the forward contract can have different characteristics to American options on the asset.

Result *The price of an American put on a forward contract equals the price of a European put, that is*

$$\widetilde{P}^{F_K}(t, T) = P^{F_K}(t, T) = P_K(t, T).$$

Therefore, we have

$$\widetilde{P}^{F_K}(t,T) \leq \widetilde{P}_K(t,T).$$

Similarly

$$\widetilde{C}^{F_K}(t,T) = C^{F_K}(t,T) = C_K(t,T).$$

American options on forward contracts are equal in value to European options on forward contracts, which are equal in value to European options on the underlying asset. American puts on the underlying asset will, however, typically have greater value.

Proof We will prove the result for puts.

$\widetilde{P}^{F_K}(t,T) \geq P^{F_K}(t,T)$ is obvious. To show $\widetilde{P}^{F_K}(t,T) \leq P^{F_K}(t,T)$, suppose the inequality does not hold, and that $\widetilde{P}^{F_K}(t,T) - P^{F_K}(t,T) = \epsilon > 0$. Then the portfolio consisting of long one European put, short one American put and a holding ϵ of cash has zero cost.

If the American put is never exercised, we do not exercise the European put, and the portfolio is worth $\epsilon/Z(t,T) > 0$ at T. If the American put is exercised early, then the portfolio now consists of long one forward contract (with delivery price K and maturity T), long one European put on the forward contract and some cash. We can exercise the European put at T to go short an identical contract. Hence the portfolio again has value $\epsilon/Z(t,T) > 0$ at T. Therefore, we have an arbitrage portfolio. ■

The contract underlying the American option—the forward contract with maturity T—and the option itself have no cashflows until T, hence the American put and call are equal to the European put and call, since no advantage can be gained by exercising early. This argument is similar to the reasoning why an American call on a stock paying no dividends has price equal to that of the European. However, the American put on a stock has price greater than or equal to the European put.

The implicit strike on our option-on-the-forward is zero. The option is to enter the forward contract with delivery price K at zero cost. We can more precisely write $C_0^{F_K}(t,T)$ for the price of this option, and we have $C_0^{F_K}(t,T) = C_K(t,T)$. Note also that

$$C_{K^*}^{F_K}(t,T) = C_{K+K^*}(t,T).$$

In words, the right to pay K^* at T to enter into a forward contract, with delivery price K and maturity T, is the same as the right to pay $K + K^*$ to buy the stock at T. The exercises provide further examples.

Example Options on FRAs—called caplets and floorlets, which we encounter further in Chapter 12—are usually all specified as European style. An American option on a FRA of fixed dates T to $T + \alpha$ would be equal in value to the European option. Options on futures are more complex because of the variation margin of the contract, and thus are not strictly an option on a forward rate. American options on futures typically have slightly higher value than European options on futures.

We revisit options on forwards in Chapter 10, where the approach provides an elegant way to price stock options in the presence of random interest rates.

7.8 EXERCISES

For Questions 1 to 4, assume all options are European style with maturity T. A 'K call' is call option with strike price K. A 'knockout' option has payout zero if the defined event occurs.

1. **Ordering option prices by monotonicity theorem I**
   Consider the following eight options I-VIII. Here, $K_1 < K_2 < K_3$.
   - I K_1 call
   - II K_1 call that knocks out if $S_T > K_2$
   - III K_1 call that knocks out if $S_t > K_2$ for any $0 \leq t \leq T$
   - IV K_1 call that knocks out if $S_T < K_1$
   - V K_1 call that knocks out if $S_t < K_1$ for any $0 \leq t \leq T$
   - VI K_1, K_2 call spread (long one K_1 call, short one K_2 call)
   - VII Digital call with strike K_1 and payout $K_2 - K_1$
   - VIII K_1, K_2, K_3 ***call ladder*** (long one K_1 call, short one K_2 call, short one K_3 call)

   For each of the pairs of A and B in Table 7.1, choose the most appropriate relationship between prices at time $t \leq T$ out of $=, \geq, \leq$ and ?, where ? means the relationship is indeterminate.

Table 7.1 Relationship of A and B.

	A	$=, \geq, \leq$ or ?	B
(a)	I		VI
(b)	II		VI
(c)	II		III
(d)	I		IV
(e)	I		V
(f)	I		VII
(g)	VI		VII
(h)	VII		VIII
(i)	III		VIII
(j)	II		VII

2. **Ordering option prices by monotonicity theorem II**
Consider the following ten options I–X. Here, $K_{i+1} = K_i + \beta$ for $i = 1, 2$, for a constant $\beta > 0$.

I K_1 call

II K_1 put

III K_3, K_2 put spread, that is, +1 K_3 put and −1 K_2 put

IV K_1, K_2, K_3 call butterfly, that is, +1 K_1 call, −2 K_2 calls and +1 K_3 call

V K_1 call that knocks out if $S_T \geq K_2$

VI Digital call with strike K_1 and payout β

VII K_1, K_3 ***digital call spread***, that is, a portfolio of +1 digital call with strike K_1 and payout β, and −1 digital call with strike K_3 and payout β

VIII Digital put with strike K_3 and payout β that knocks out if $S_T \leq K_1$

IX A portfolio of +1 K_1 call and −2 K_2 calls, all of which knock out if $S_t < K_3$ for any $0 \leq t \leq T$

X A portfolio of +1 K_1 call and −2 K_2 calls, all of which knock out if $S_T > K_3$

For each of the pairs of A and B in Table 7.2, choose the most appropriate relationship between prices at time $t \leq T$ out of =, ≥, ≤ and ?, where ? means the relationship is indeterminate.

Table 7.2 Relationship of A and B.

	A	=, ≥, ≤ or ?	B
(a)	I		II
(b)	II		III
(c)	III		IV
(d)	IV		V
(e)	IV		X
(f)	VII		VIII
(g)	III		VIII
(h)	IV		VII
(i)	II		IX
(j)	II		X

3. **Butterflies, condors and call ladders**

(a) Recall that a call butterfly with strikes $(K_1, K_1 + \beta, K_1 + 2\beta)$, for some fixed $\beta > 0$, is a portfolio consisting of +1 K_1 call, +1 $(K_1 + 2\beta)$ call and −2 $(K_1 + \beta)$

calls. Using put-call parity or otherwise, restate the call butterfly as a portfolio consisting solely of puts.

(b) A call ***condor*** is a portfolio consisting of $+1$ K call, -1 $(K + \beta)$ call, -1 $(K + 2\beta)$ call and $+1$ $(K + 3\beta)$ call. Draw the payout of the condor, and express the condor as a portfolio consisting solely of call butterflies.

(c) A call ***ladder*** consists of $+1$ K call, -1 $(K + \beta)$ call and -1 $(K + 2\beta)$ call. What relationships hold between the prices at time $t \leq T$ of the call ladder, butterfly and condor with common maturity T?

4. **Call spreads and digital options**

(a) Draw the payout profile for the following two call spread portfolios.

(i) $+1$ K call and -1 $(K + 1)$ call.

(ii) $+2$ K calls and -2 $(K + 0.5)$ calls.

(b) By constructing a series of portfolios of call spreads and taking limits, prove that the price at time t of a digital call, with strike K^* and payout 1, is given by

$$-\frac{\partial}{\partial K} C_K (t, T) \mid_{K^*},$$

where $|_{K^*}$ means the function is evaluated at $K = K^*$.

(c) Write down the equivalent formula for a digital put option in terms of put prices.

(d) By examining the payout profile, derive a put-call parity relationship for the digital call and digital put.

5. **Bounds on European and American puts**

(a) Using put-call parity and bounds on a European call, or otherwise, prove that the price of a European put on a stock paying no dividends satisfies

$$\max \{0, KZ(t, T) - S_t\} \leq P_K(t, T) \leq KZ(t, T).$$

(b) Prove that the price of an American put on a non-dividend paying stock satisfies

$$\max \{0, K - S_t\} \leq \tilde{P}_K(t, T) \leq K.$$

(c) Give an example to show that the American put can be worth more than the European put, that is, where immediate exercise of the American gives payout at time T strictly greater than the payout of the European.

Hint Exercising early means the strike price K will be received early. Consider under which environments this will be preferable.

6. **Options with different exercise dates**
   (a) By considering a portfolio of ZCBs and a call option, prove that the price at time $t \leq T$ of a call option with strike K on a stock that pays no dividends satisfies

$$C_K(t,T) \geq \max\{S_t - KZ(t,T), 0\}.$$

   (b) Hence prove that if $t \leq T_1 \leq T_2$,

$$C_K(t,T_2) \geq C_K(t,T_1).$$

   Hint Consider the case $t = T_1$.

   (c) Does the same result hold for puts? That is, prove or find a counter example to the statement

$$P_K(t,T_2) \geq P_K(t,T_1) \text{ for } t \leq T_1 \leq T_2.$$

7. **Options on forwards**
   Recall that $C_{K^*}^{F_K}(t,T)$ is the price at current time t of a European call, with maturity T and strike K^*, on a forward contract with delivery price K and same maturity T; that is, the right to pay K^* at time T to go long a forward with delivery price K and maturity T. Furthermore, $P_{K^*}^{F_K}(t,T)$ is the value of the European put on the forward contract; that is, the right to receive K^* to go short the forward contract with delivery price K. $\tilde{C}_{K^*}^{F_K}(t,T)$ and $\tilde{P}_{K^*}^{F_K}(t,T)$ are the American-style versions of these options on forward contracts.
   (a) Which of the following assertions are true? Prove, or find a counter example to, each.
      (i) $P_{K^*}^{F_K}(t,T) = P_{K+K^*}(t,T)$
      (ii) $C_{K^*}^{F_K}(t,T) = C_{K+K^*}(t,T)$
      (iii) $\tilde{P}_{K^*}^{F_K}(t,T) = P_{K+K^*}(t,T)$
      (iv) $\tilde{C}_{K^*}^{F_K}(t,T) = C_{K+K^*}(t,T)$
   (b) Prove that $\tilde{P}_{K^*}^{F_K}(t,T) \leq \tilde{P}_{K+K^*}(t,T)$.

   Hint For each put option consider when the amounts K and K^* may be received.

PART III

Replication, Risk-neutrality and the Fundamental Theorem

8

Replication and risk-neutrality on the binomial tree

We now develop important concepts at the heart of quantitative finance, in particular the links between replication arguments, no-arbitrage and ***risk-neutral probabilities***. A simple model, the one-step binomial tree, is rich enough to introduce and explore these key ideas.

8.1 Hedging and replication in the two-state world

Consider a one-step binomial (that is, two-state) world as shown in Figure 8.1, where a stock that pays no dividends has price 100 at time $T = 0$, and at time $T = 1$ can either increase by 20% to 120 (state A), or decrease by 10% to 90 (state B).

Assume annually compounded interest rates are 10%, and consider a one-year call option with strike 110. Its payout at $T = 1$ in state A is 10 and its payout in state B is 0.

Consider a portfolio consisting of short one 110-strike call and long Δ units of stock. At $T = 1$, the portfolio value in state A is $-10 + 120\Delta$, and in state B is 90Δ. These are the same if $\Delta = 1/3$.

Therefore, the portfolio consisting of short one call option and long 1/3 stock, and the portfolio consisting of 30 zero coupon bonds with maturity $T = 1$, have the same value in

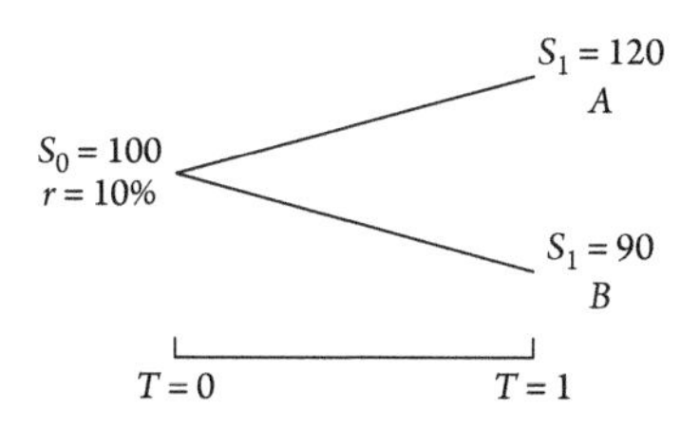

Figure 8.1 One-step binomial tree

all states of the world at $T = 1$. By the corollary to the monotonicity theorem, they have the same value at time $T = 0$. Therefore,

$$\frac{100}{3} - C_{110}(0,1) = \frac{30}{1.1}$$

and so

$$C_{110}(0,1) = \frac{200}{33}.$$

We have thus priced the call option without any assumption about the probability of either state. This reasoning is sometimes called a hedging argument, since we have constructed a portfolio of the call option and stock that has no risk—is hedged—to movements in the stock price.

A complementary approach is to consider a replication argument. In a two-state world there must be a unique linear combination of the stock and zero coupon bond that agrees with the option value in both states at time $T = 1$, that is, a portfolio of λ stocks and μ ZCBs which matches the option payout at $T = 1$. In our example λ and μ satisfy the equations

$$\begin{cases} 120\lambda + \mu = 10 \\ 90\lambda + \mu = 0 \end{cases}$$

with solution $\lambda = 1/3$ and $\mu = -30$. Thus the replicating portfolio is long 1/3 stock and short 30 bonds, and by the corollary to the monotonicity theorem the price of the option is the value of this portfolio at time $T = 0$. So we have

$$C_{110}(0,1) = \frac{100}{3} - \frac{30}{1.1} = \frac{200}{33}.$$

We can think of this replication graphically. In the two-state world, a derivative of the stock with payout $g(S_T)$ at time T can be represented as two points in the $(S_T, g(S_T))$ plane. There is a (unique) line connecting these points. Its intercept at $S_T = 0$ is the number of ZCBs in the replicating portfolio, and the slope is the number of stocks.

The hedging argument constructs a portfolio of stock and option that is hedged to any movement in the stock price, that is, it uses the option and stock to replicate the bond. The replication argument constructs a portfolio of stock and bond to replicate the option. We could equally use a portfolio of option and bond to replicate the stock. There are two states of the world and thus any two linearly independent assets will span a payout at $T = 1$.

If the option price is different to the price given by the replication argument, then we can construct an arbitrage portfolio. For example, if for some $\epsilon > 0$ the option price is ϵ higher than the replicating price, then the portfolio consisting of short one option, long the replicating portfolio and a holding of ϵ cash is an arbitrage portfolio. So the replication argument determines the only possible arbitrage-free price, though we have not yet shown that this price is in fact arbitrage-free.

More generally, for a derivative with payout $D(\omega_A)$ in state A and $D(\omega_B)$ in state B, we can find the replicating portfolio by solving the linear equations

$$\lambda S_1(\omega_A) + \mu = D(\omega_A) \text{ and } \lambda S_1(\omega_B) + \mu = D(\omega_B).$$

We explore this general case in the exercises.

8.2 Risk-neutral probabilities

Suppose in the example above, state A has probability p and state B has probability $(1 - p)$, for $0 < p < 1$. (Note if $p = 1$ or $p = 0$ there is an immediate arbitrage. You may want to write down the relevant arbitrage portfolios in these cases.) Then the present value of the expected value of the option payout is given by

$$Z(0,1)E(S_1 - K)^+ = \frac{1}{1.1}\big(10p + 0(1 - p)\big) = \frac{10p}{1.1}.$$

The unique value of p for which this expectation equals the only possible arbitrage-free price is p^* such that

$$\frac{10p^*}{1.1} = \frac{200}{33}, \text{ hence } p^* = \frac{2}{3}.$$

Now consider the expected stock price

$$E(S_1) = 120p + 90(1 - p).$$

Putting $p = p^* = 2/3$ we have

$$E_*(S_1) = 110 = 100(1 + r)$$

where E_* is the expectation using probability p^* for state A.

Thus we have a remarkable result. Using the probability p^*, under which the present value of the expected option payout is the only possible arbitrage-free option price (found by the replication argument), the expected value of the stock price is its forward price (and its expected return is the interest rate r). The value of p^* for which this holds is called the ***risk-neutral probability***.

We have shown that in this particular one-step two-state tree, the only possible arbitrage-free option price is the price given by ***risk-neutral pricing***, that is, where prices are discounted expectations of payouts using the risk-neutral probability.

There is nothing special about the numbers we used in this example. The result holds in general for an arbitrary one-step binomial tree and a general derivative contract maturing at $T = 1$. We prove the general case in Question 3.

The terminology 'risk-neutral probability' captures the fact that these probabilities depend on the values the stock can take in the two states of the world, but not on the actual probability (or risk) of those states.

Note that we have

$$Z(0,1) = Z(0,1)E_*(Z(1,1)), \text{ a tautological result since } Z(1,1) = 1,$$

$$S_0 = Z(0,1)E_*(S_1)$$

and

$$C_K(0,1) = Z(0,1)E_*(C_K(1,1)).$$

In other words, risk-neutral pricing holds for all the assets. The price of any contract is the discounted risk-neutral expectation of its payout at $T = 1$.

By linearity of expectation, this must also be true for any linear combination of assets. In particular, if a portfolio has zero value at $T = 0$, the expected value of its payout at maturity $T = 1$ is zero. So, either the portfolio is identically zero at maturity or, if it has positive value at maturity with positive probability, then it must also have negative value at maturity with positive probability. Therefore, it cannot be an arbitrage portfolio. Thus for the one-step binomial tree, if prices of contracts are discounted risk-neutral expected values, there can be no arbitrage.

This argument holds regardless of the actual probability of an up move on the tree, since arbitrage is only defined using zero or non-zero probabilities. If arbitrage exists for some value of p, $0 < p < 1$, it exists for all p. Similarly if no arbitrage exists for some $0 < p < 1$, none exists for all p. So for the one-step binomial model, we have shown that the risk-neutral price is arbitrage-free.

Summary The replication and risk-neutral pricing arguments we have explored in these two sections provide a powerful result. The replication argument determines that only one price can be free of arbitrage and that any other price is arbitrageable. Risk-neutral pricing demonstrates that this replication price is indeed arbitrage-free. Therefore, we have obtained a unique arbitrage-free price.

We extend these key arguments underlying quantitative finance to multiple time steps in the next section. The arguments can also be extended to continuous-time models, although this theory is largely beyond our scope. However, in Chapter 10 we take the limit of the binomial tree and assume the arguments we developed in discrete time transfer to the continuous case. We do not discuss the mathematical conditions required.

8.3 Multiple time steps

Suppose we now have two binomial steps, as shown in Figure 8.2. We assume that the stock increases 20% or decreases 10% at each step, and annually compounded interest rates are a constant 10%. We could vary these assumptions at each step without altering our conclusions. However, the arithmetic would become messier without increasing insight.

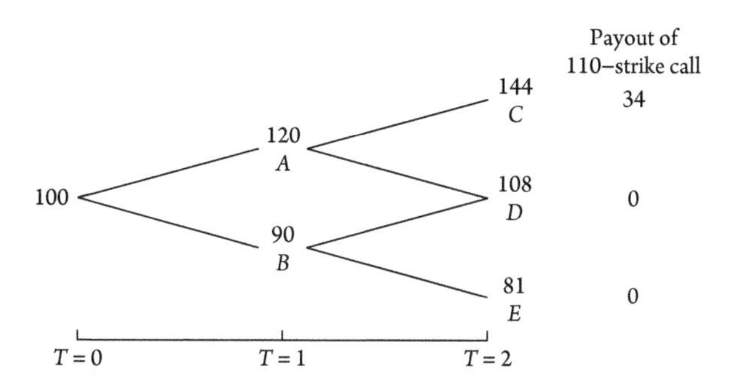

Figure 8.2 Two-step binomial tree

Consider the value of a two-year 110-strike call. We will apply similar replication arguments as before at states A and B.

The option value at state B is trivially zero. At state A, the replicating portfolio of λ_A stocks and μ_A ZCBs with maturity $T = 2$ satisfies

$$144\lambda_A + \mu_A = 34 \text{ and } 108\lambda_A + \mu_A = 0.$$

So

$$\lambda_A = \frac{34}{36} \text{ and } \mu_A = -102$$

and the option price at state A is given by

$$120\left(\frac{34}{36}\right) - 102\left(\frac{1}{1.1}\right) = \frac{680}{33}.$$

Therefore, the random variable $C_{110}(1, 2)$ has value $680/33$ in state A and 0 in state B.

We can then set up a replicating portfolio at time $T = 0$ of λ_0 stocks and μ_0 ZCBs with maturity $T = 1$, where we find that

$$\lambda_0 = \frac{68}{99} \text{ and } \mu_0 = -\frac{68}{1.1}.$$

The price at $T = 0$ of the two-year 110-strike call is thus

$$100\left(\frac{68}{99}\right) - \frac{68}{(1.1)^2} = \left(\frac{4}{9}\right)\left(\frac{34}{(1.1)^2}\right).$$

The calculation of the option price is easier via risk-neutral pricing where p^*, the probability of an up move, is $2/3$ at each node. Risk-neutral probabilities are $(2/3)^2, 2(2/3)(1/3)$ and $(1/3)^2$ for states C, D and E respectively, and thus the option price given by risk-neutral pricing is

$$C_{110}(0,2) = Z(0,2)E_*(S_2 - K)^+ = \frac{1}{(1.1)^2}\left(\frac{2}{3}\right)^2 34,$$

identical to that given by replication. We also see that

$$E_*(S_2) = \left(\frac{2}{3}\right)^2 144 + 2\left(\frac{2}{3}\right)\left(\frac{1}{3}\right) 108 + \left(\frac{1}{3}\right)^2 81 = 121 = 100(1+r)^2.$$

The option price obtained by replication equals the risk-neutral price, that is, the discounted risk-neutral expectation of its payout. Under the risk-neutral probabilities, we also have $E_*(S_2 \mid S_0) = S_0(1+r)^2$.

At each node of the tree the stock could only move to two possible values, hence exact replication is possible. Replication and risk-neutral pricing immediately extend to multiple binomial steps.

8.4 General no-arbitrage condition

Consider a general version of the binomial tree shown in Figure 8.3, where at each node the stock can move from price S_{n-1} either to price $S_n = S_{n-1}(1+u)$ or $S_n = S_{n-1}(1+d)$, where $d < u$. We assume the probability of an up move and a down move are both non-zero. Let the constant interest rate be r, and the size of each time step be $\Delta T = 1$. (In Chapter 10 we will let ΔT become small.)

Theorem *The binomial tree is arbitrage-free* $\Longleftrightarrow d < r < u$.

Proof To prove the right implication, we suppose $r \geq u$ and demonstrate that we can construct an arbitrage portfolio. In particular, we sell one stock at time $T = 0$ for S_0 and invest proceeds at r. At time $T = 1$ we have $S_0(1+r)$ of cash, and the short stock position is worth either $S_0(1+u) \leq S_0(1+r)$ or $S_0(1+d) < S_0(1+r)$. Therefore, this is an arbitrage portfolio since the probability of the stock price equalling $S_0(1+d)$ is non-zero.

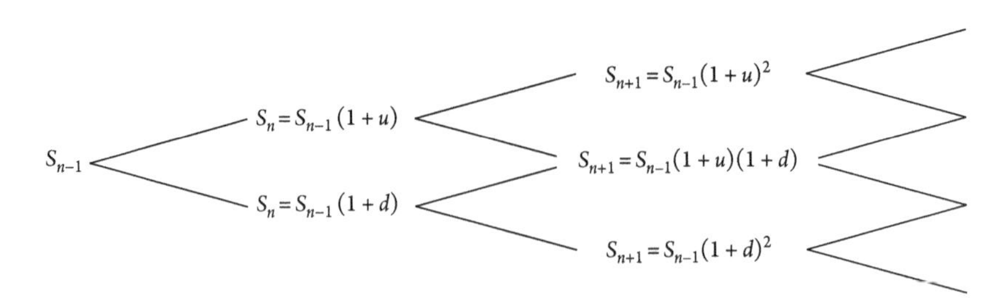

Figure 8.3 Binomial tree

A similar argument applies if $r \leq d$. We construct an arbitrage portfolio by borrowing S_0 and buying one stock.

To prove the left implication, consider all portfolios consisting of λ stocks and μ bonds, with $\lambda, \mu \in \mathbb{R}$, which have zero value at $T = 0$. We need to show that all such portfolios are arbitrage-free.

The value of the portfolio at $T = 0$ is $\lambda S_0 + \frac{\mu}{1+r}$, and so portfolios with zero value must be of the form

$$(\lambda, \mu) = \left(\frac{\nu}{S_0}, -\nu(1+r) \right), \text{ for } \nu \in \mathbb{R}.$$

Write $V(1)$ for the value of the portfolio at time $T = 1$.

(a) If $\nu = 0$, then trivially $V(1) = 0$.

(b) If $\nu > 0$, then if the stock goes down

$$V(1) = \nu \frac{S_0(1+d)}{S_0} - \nu(1+r) < 0.$$

(c) If $\nu < 0$, then if the stock goes up

$$V(1) = \nu \frac{S_0(1+u)}{S_0} - \nu(1+r) < 0.$$

Therefore, there can be no arbitrage portfolios since there is always a non-zero probability of the portfolio having negative value at $T = 1$. Note that it is important for this argument that both states have non-zero probabilities. ■

Suppose the probability of an up move on each node is $p > 0$, then

$$\begin{aligned} E(S_1 \mid S_0) &= S_0 \left(p(1+u) + (1-p)(1+d) \right) = S_0 \left(1 + pu + (1-p)d \right) \\ &= S_0(1+r) \iff r = pu + (1-p)d. \end{aligned}$$

Therefore, the risk-neutral probability p^* is given by

$$p^* = \frac{r-d}{u-d}.$$

We have thus established the following equivalent conditions.

Result *No arbitrage portfolios* $\iff d < r < u \iff 0 < p^* < 1$.

That is, for the binomial tree, the absence of arbitrage portfolios is equivalent to the existence of a unique risk-neutral probability p^* under which prices are discounted expected values of the payout at maturity.

We can easily show that the replicating portfolio for a general derivative contract, with payout γ in the down state and $\gamma + \beta$ in the up state, for $\gamma, \beta \in \mathbb{R}$, is λ stocks and μ ZCBs, where

$$\lambda = \frac{\beta}{S_0(u-d)} \text{ and } \mu = \gamma - \frac{\beta(1+d)}{(u-d)}.$$

Therefore, the replication price is given by

$$\frac{\beta}{(u-d)} + \left(\gamma - \frac{\beta(1+d)}{(u-d)}\right)\frac{1}{1+r} = \frac{1}{1+r}\left(\gamma + \frac{(r-d)\beta}{(u-d)}\right).$$

This can be re-expressed as

$$\frac{1}{1+r}\left(p^*(\gamma+\beta) + (1-p^*)\gamma\right),$$

the discounted risk-neutral expectation. Thus we have shown the equivalence of the replication and risk-neutral price in the general case.

When we have n identical time steps, the risk-neutral probability distribution for S_n is binomial with

$$P^*\left(S_n = S_0(1+u)^j(1+d)^{n-j}\right) = \binom{n}{j} p^{*j}(1-p^*)^{n-j} \text{ for } j = 0, \ldots, n.$$

For a derivative with payout $g(S_n)$ at time n, its price at time $T = 0$ is, therefore, given by

$$\frac{1}{(1+r)^n}E_*(g(S_n)) = \frac{1}{(1+r)^n}\sum_{j=0}^{n}\binom{n}{j}p^{*j}(1-p^*)^{n-j}g\left(S_0(1+u)^j(1+d)^{n-j}\right).$$

For a call option, $g(S_n) = (S_n - K)^+$, and we can simplify this expression to some degree to obtain the Cox Ross Rubinstein formula (see Question 4). However, it is generally easier to take limits and work with continuous distributions. We do this in Chapter 10.

8.5 EXERCISES

1. **Binomial tree: European and American puts**
   Consider a two-step binomial tree, where a stock that pays no dividends has current price 100, and at each time step can increase by 20% or decrease by 10%. The possible values at time $T = 2$ are thus 144, 108 and 81. The annually compounded interest rate is 10%.

(a) Calculate the price of a two-year 106-strike European put using (i) a replication argument and (ii) risk-neutral expectation.

(b) Calculate the price of a two-year 106-strike American put using replication, and hence verify that the American put has price strictly greater than the European.

(c) Calculate the prices of a two-year 86-strike European put and American put. What is different to (b)?

2. **Arbitrage on the tree**
A stock that pays no dividends has price today of 100. In one year's time the stock is worth 110 with probability 0.75, and 85 with probability 0.25. The one-year annually compounded interest rate is 5%.

(a) Calculate the forward price of the stock for a forward contract with maturity one year.

(b) Calculate the price of a one-year European put option with strike 100.

(c) Suppose you observe that the put option in part (b) has a market price of 4. Determine an arbitrage portfolio and calculate how much profit is generated at time $T = 1$ by this portfolio.

3. **General two-state world**
A non-dividend paying stock has current price S_0. In one year's time there are two possible states of the world. The stock may be worth S_U in state A, or S_D in state B, with $S_U > S_D$. A derivative contract pays $\gamma + \beta$ in state A and γ in state B, for some $\gamma, \beta > 0$. The annually compounded interest rate is a constant r.

(a) What is the forward price for the stock?

(b) Prove that the portfolio consisting of short one derivative and long Δ stocks is hedged—that is, has the same payout in state A and state B—if

$$\Delta = \frac{\beta}{S_U - S_D}.$$

Hence prove by replication that the price of the derivative is equal to

$$\frac{1}{(1+r)}\left(\gamma + \beta\left(\frac{S_0(1+r) - S_D}{S_U - S_D}\right)\right).$$

(c) Find the risk-neutral probability p^* for state A. Hence calculate the discounted expected payout of the derivative under risk-neutral probabilities, and verify that your answer equals the replication price in (b).

(d) What relationships must hold between S_U, S_D, S_0 and r for the risk-neutral probability to exist? What have we assumed about the actual probabilities of state A and state B?

(e) If $\beta = \gamma(S_U/S_D - 1)$, what is the derivative contract? Verify the price (b) makes sense.

(f) If $\beta = 0$, what is the derivative contract? Again, verify the price (b) makes sense.

4. **Cox Ross Rubinstein formula**
   In the n-step binomial tree, the discounted risk-neutral expectation of the option payout is given by

$$\frac{1}{(1+r)^n}\sum_{j=0}^{n}\binom{n}{j}p^{*j}(1-p^*)^{n-j}g\left(S_0(1+u)^j(1+d)^{n-j}\right).$$

Consider a European call option with maturity n where $g(S_n) = (S_n - K)^+$. By choosing m to be the least number such that

$$S_0(1+u)^m(1+d)^{n-m} > K,$$

show that the call option price is given by

$$S_0\,B(n,a,m) - \frac{K}{(1+r)^n}B(n,p^*,m)$$

where $B(n,p,k)$ is defined to be equal to $P(X \geq k)$ with $X \sim$ binomial(n,p), and

$$a = \frac{p^*(1+u)}{(1+r)}.$$

Hint Use the fact that

$$(1-a) = \frac{(1-p^*)(1+d)}{(1+r)}.$$

9

Martingales, numeraires and the fundamental theorem

In this chapter we tie together our work on replication and risk-neutrality with the probabilistic concept of a martingale, to establish the fundamental theorem of asset pricing. The fundamental theorem precisely establishes the link between expected values and arbitrage-free prices.

9.1 Definition of martingales

A series of random variables $X_0, \ldots, X_n, \ldots$ is a ***martingale*** if

$$E(| X_i |) < \infty \text{ for all } i, \text{ and } E(X_{n+1} \mid X_0, \ldots, X_n) = X_n.$$

If $X_0, \ldots, X_n$ is ***Markov***—that is, $P(X_{n+1} \mid X_0, \ldots, X_n) = P(X_{n+1} \mid X_n)$, or X_n captures all the information about the conditional distribution of X_{n+1}—then the martingale condition becomes

$$E(X_{n+1} \mid X_n) = X_n.$$

All processes we consider in this book are assumed to be Markov. By the tower law of iterated expectations,

$$E(X_{n+1} \mid X_0) = E(E(X_{n+1} \mid X_n) \mid X_0).$$

Therefore, the martingale condition gives

$$E(X_{n+1} \mid X_0) = E(X_n \mid X_0) = \ldots = X_0$$

and

$$E(X_n \mid X_m) = X_m \text{ for any } m \leq n.$$

A martingale is the probabilistic definition of a 'fair game', where the expected value of a portfolio's value tomorrow is its value today.

X_n is a martingale ***with respect to*** $Y_0, \ldots, Y_n$ if

$$E(| X_i |) < \infty \text{ for all } i, \text{ and } E(X_{n+1} \mid Y_n) = X_n.$$

The process $Y_0, \ldots, Y_n$ is called a ***filtration***. One can think of this as the price information known at time n. In finance often $Y_n = S_n$, the stock price, and we see shortly that the martingale X_n is often the discounted price of a derivative of the stock. Martingales play an important role in probability theory and finance (see the excellent Williams (1991) for a more detailed exposition).

Result *On the binomial tree* $E_*(S_n \mid S_0) = S_0(1 + r)^n$.

Proof The statement holds for $n = 1$. We assume it holds for $n - 1$ and prove by induction. We have

$$E_* (S_n \mid S_{n-1} = y) = y(1 + u)p^* + y(1 + d)(1 - p^*) = y(1 + r)$$

and thus

$$E_* (S_n \mid S_{n-1}) = S_{n-1}(1 + r).$$

Again by the law of iterated expectations

$$E_*(S_n \mid S_0) = E_* (E_* (S_n \mid S_{n-1}) \mid S_0) = (1 + r)E_* (S_{n-1} \mid S_0).$$

We have proof by induction. ∎

More generally, we have

$$E_*(S_n \mid S_m) = S_m(1 + r)^{(n-m)}, \text{ for } m \leq n.$$

Recall that in Chapter 1 we defined M_n to be the value of the money market account. In particular, $M_0 = 1$ and, if r is the constant annually compounded interest rate, $M_n = (1 + r)^n$. We can rewrite our general result as

$$E_* \left(\frac{S_n}{M_n} \mid S_m \right) = \frac{S_m}{M_m}, \text{ for } m \leq n.$$

Therefore, we immediately have the following.

Result *On the binomial tree,* S_m/M_m*, the discounted stock price, is a martingale under the risk-neutral probability.*

Now note that the price of a call with maturity date n satisfies

$$C_K(0, n) = \frac{1}{(1 + r)^n} E_* ((S_n - K)^+ \mid S_0).$$

In general we have

$$\frac{C_K(m,n)}{M_m} = E_* \left(\frac{C_K(n,n)}{M_n} \mid S_m \right),$$

and have established the following.

Result *On the binomial tree, the discounted call price $C_K(m,n)/M_m$ is a martingale under the risk-neutral probability.*

We see that risk-neutral pricing is equivalent to the statement that ratios of prices to the money market account are martingales under the risk-neutral probability.

9.2 Numeraires and fundamental theorem

We now aggregate our results of the past two chapters into the following powerful result.

Fundamental theorem of asset pricing. *There are no arbitrage portfolios $\Longleftrightarrow$ there exists a risk-neutral probability distribution P^* such that the ratios of asset prices to the money market account are martingales under P^*.*

We have proved this result for the binomial tree and shown that the probability distribution P^* (determined by the binomial probability p^* at each node) is unique. In particular, when interest rates are a constant r, and $M_n = (1+r)^n$, the fundamental theorem on the binomial tree states that there are no arbitrage portfolios $\Longleftrightarrow$ derivative prices are discounted expected values of their payouts at maturity.

There are two important extensions to the fundamental theorem.

First, under suitable conditions, the result holds for continuous-time models. We will not attempt to prove this extension, although we will take the limit of the binomial tree as the number of time steps tends to infinity and assume results carry over to this limit. The general theorem in continuous time, a landmark result, was proven by Harrison and Kreps (1979).

Secondly, the choice of the money market account as the unit by which we discount or rebase prices is unimportant. We call the rebasing unit the ***numeraire***. In the case of the money market account, the numeraire can be thought of as 'the price of time-t money', that is, how much one unit of cash is worth at time t. However, there is nothing special about the money market account. Any positive asset can be used as a numeraire. The intuition behind the fundamental theorem is that no-arbitrage is equivalent to the ratio of two assets remaining the same (in risk-neutral expected value terms) over time—loosely speaking, one asset cannot on average grow faster than another. Provided we have two linearly independent assets (for example, the ZCB and the stock) then we can eliminate risk at each step of the binomial tree, and risk-neutral and martingale arguments apply.

We restate a more general version of the fundamental theorem, which we do not prove here.

Theorem *There are no arbitrage portfolios* $\Longleftrightarrow$ *for a given positive asset (with price* $N_t > 0$ *at time* t*) there exists a probability* Q^**, defined over the same set of possible outcomes, such that the ratios of asset prices to the numeraire* N_t *are martingales under* Q^**.*

By 'same set of possible outcomes' we mean that the set of outcomes which have positive probability are the same under Q^* as under the actual probability distribution. In other words, the null sets of both probability distributions are the same; such probability distributions are called ***equivalent*** distributions.

The probability distribution Q^* will be different depending on the choice of the numeraire. We call Q^* the ***risk-neutral distribution with respect to*** N_t. If the numeraire is the money market account, we often call Q^* simply the risk-neutral distribution (dropping explicit mention of a numeraire). We explore the change of numeraire in the next section, and in the exercises.

The fundamental theorem means that, for a derivative of S_t with maturity T and price $D(t, T)$ at time $t \leq T$, no-arbitrage is equivalent to the statement

$$\frac{D(t,T)}{N_t} = E_*\left(\frac{D(T,T)}{N_T} \mid S_t\right),$$

where E_* is the expectation under the risk-neutral distribution with respect to N_t.

So far, we have implicitly used the money market account as numeraire, which has the nice property $M_0 = 1$. But we have also been assuming up to this point that r is a constant and $M_T = e^{rT}$ is deterministic, so we could take it in or out of the expected value. In general, however, r and hence M_T will not be constant. We can navigate the complexity of random interest rates in an elegant manner by choosing an alternative numeraire, the ZCB with maturity T. Now $N_t = Z(t, T)$, which has the nice property $Z(T, T) = 1$. This holds regardless of whether interest rates are constant. Note that this choice of numeraire depends upon T, the maturity of the derivative. Then we obtain

$$D(t,T) = Z(t,T)E_*(D(T,T) \mid S_t),$$

where E_* now is the risk-neutral expectation with respect to the ZCB numeraire. This expression, again a discounted expected value, holds even if interest rates are random, and becomes especially useful when we consider interest rate options in Part IV, when interest rates certainly cannot be assumed to be constant.

Note that the forward contract with value $V_K(t, T)$ is itself a derivative contract, and thus by the fundamental theorem $V_K(t, T)/Z(t, T)$ must be a martingale. Therefore,

$$\frac{V_K(t,T)}{Z(t,T)} = E_*\left(\frac{V_K(T,T)}{Z(T,T)} \mid S_t\right) = E_*\left((S_T - K) \mid S_t\right) = E_*(S_T \mid S_t) - K.$$

The fundamental theorem applied to the stock itself asserts that the ratio $S_t/Z(t, T)$ must also be a martingale under E_*, and so

$$\frac{S_t}{Z(t,T)} = E_*\left(S_T \mid S_t\right).$$

Combining these two results shows that the forward price $F(t, T)$, the value of K such that $V_K(t, T) = 0$, is given by

$$F(t, T) = E_* (S_T \mid S_t) = \frac{S_t}{Z(t, T)}.$$

We have here derived the forward price for the stock simply using the fundamental theorem, without making any assumptions about the risk-neutral distribution or whether interest rates are constant. We had to be able to accomplish this, since we earlier derived the forward price directly from the current stock and ZCB prices without making any assumptions about its distribution, or even about possible values for the stock price. It is reassuring that the two approaches (the link being the proof of the fundamental theorem) come up with the same result. We also have now, for the first time, identified the forward price as an expected value, in particular the expected value of the stock price under the risk-neutral distribution with respect to the zero coupon bond numeraire.

9.3 Change of numeraire on binomial tree

The one-step binomial tree again provides great insight as to how the risk-neutral distribution changes, as we change from one numeraire to another. We present the following worked example.

Recall the tree in Figure 8.1, where $S_0 = 100$ and $Z(0, 1) = 1/1.1$. At time $T = 1$, $S_1 = 120$ or 90. Consider a one-year 110-strike call, whose price at time m we denote $C_{110}(m, 1)$ for $m = 0, 1$. Using the money market account as the numeraire, S_m/M_m and $C_{110}(m, 1)/M_m$ must both be martingales under the risk-neutral distribution $p^* = 2/3$. In particular,

$$\frac{C_{110}(0, 1)}{1} = E_{p^*}\left(\frac{C_{110}(1, 1)}{M_1}\right) = \left(\frac{2}{3}\right)\left(\frac{10}{1.1}\right) = \frac{200}{33},$$

where the notation E_{p^*} denotes the risk-neutral expectation with respect to the money market numeraire.

We now change to the stock numeraire. By the fundamental theorem, $Z(m, 1)/S_m$ must be a martingale under the risk-neutral probability with respect to the stock numeraire. In particular, if q^* is the risk-neutral probability with respect to the stock numeraire of state A, then we require

$$\left(\frac{1}{1.1}\right)\left(\frac{1}{100}\right) = \frac{q^*}{120} + \frac{(1 - q^*)}{90} \Rightarrow q^* = \frac{8}{11}.$$

Equivalently,

$$\frac{Z(0, 1)}{S_0} = E_{q^*}\left(\frac{Z(1, 1)}{S_1}\right),$$

where the expectation E_{q^*} is the risk-neutral expectation with respect to the stock numeraire. Now, $C_{110}(m, 1)/S_m$ also must be a martingale under q^*, so

$$\frac{C_{110}(0,1)}{100} = E_{q^*}\left(\frac{C_{110}(1,1)}{S_1}\right) = \left(\frac{8}{11}\right)\left(\frac{10}{120}\right) \Rightarrow C_{110}(0,1) = \frac{200}{33}.$$

We have provided a simple illustration of a deep concept. We have two equivalent pairs (M_m, p^*) and (S_m, q^*) of numeraire and risk-neutral probabilities, either of which we can use to price derivatives consistently. The change of probability from p^* to q^* is sometimes called a ***change of measure***. We found the probabilities by invoking martingale conditions. Prices of derivative contracts are the same under either pair of numeraire and risk-neutral probability. The only facts we used about the real world probabilities of the up and down states are that only two states of the world are possible at each node, and that both states have non-zero probability.

The continuous-time analogue of the change of measure involves the Radon-Nikodym derivative and Girsanov's theorem (see Hunt and Kennedy, 2004).

9.4 Fundamental theorem: a pragmatic example

Before we take the continuous limit, we will give a pragmatic, approximate example from continuous time that shows how the fundamental theorem leads to an intuitive form for derivative prices. By the fundamental theorem, we know that the call price

$$C_K(t,T) = Z(t,T)E_*\left((S_T - K)^+ \mid S_t\right)$$

where E_* is the risk-neutral expectation with respect to the $Z(t, T)$ numeraire. We also know that the stock price satisfies

$$\frac{S_t}{Z(t,T)} = E_*\left(\frac{S_T}{Z(T,T)} \mid S_t\right).$$

Previously we determined the risk-neutral distribution uniquely from the possible states of the world on the binomial tree. We assume here without proof that an analogous process can be followed in continuous time. Consider a possible risk-neutral distribution of the form $S_T \mid S_t \sim N(\mu, \psi^2(T-t))$. Whilst we immediately note that this allows non-zero probability of negative stock prices, let us assume for the moment that this probability is negligibly small and that the normal distribution is a reasonable approximation to the risk-neutral distribution. In practice, approximate approaches like this can often be useful. Then we immediately have

$$\mu = \frac{S_t}{Z(t,T)},$$

the forward price $F(t,T)$. Writing $F = F(t,T)$ we thus have

$$C_K(t,T) = Z(t,T) \int_{-\infty}^{\infty} \frac{1}{\sqrt{2\pi}\,\psi\sqrt{T-t}} e^{-\frac{(x-F)^2}{2\psi^2(T-t)}} (x-K)^+ \, dx.$$

This integral is familiar from the exercises in the preface. Change limits and use $x - K = (x - F) + (F - K)$ to obtain

$$C_K(t,T) = Z(t,T) \left(\frac{\psi\sqrt{T-t}}{\sqrt{2\pi}} e^{-\frac{(K-F)^2}{2\psi^2(T-t)}} + (F-K)\,\Phi\left(\frac{F-K}{\psi\sqrt{T-t}}\right)\right)$$

where $\Phi(\cdot)$ is the normal cumulative distribution function.

Using put-call parity, it is straightforward to obtain put and straddle prices. Setting $k = \frac{F-K}{\psi\sqrt{T-t}}$ (a normalized measure of how far in-the-money-forward the option is), we find that the straddle price equals

$$Z(t,T)\left(\sqrt{\frac{2}{\pi}}\psi\sqrt{T-t}\,e^{-\frac{k^2}{2}} + k\psi\sqrt{T-t}(2\Phi(k)-1)\right).$$

Using Taylor series expansions for $\Phi(\cdot)$ and the exponential function, we show (see Question 7 in the exercises) that the straddle price is

$$Z(t,T)\sqrt{\frac{2}{\pi}}\psi\sqrt{T-t}\left(1 + \frac{k^2}{2} + \text{higher order terms}\right).$$

For the at-the-money-forward (ATMF) straddle—that is, when $K = F$—the price simplifies to

$$Z(t,T)\sqrt{\frac{2}{\pi}}\psi\sqrt{T-t} \approx 0.8\ \ Z(t,T)\psi\sqrt{T-t}.$$

This is a handy approximation for quick calculation of ATMF straddle prices.

9.5 Fundamental theorem: summary

Using the numeraire $Z(t,T)$, the fundamental theorem allows us to write $D(t,T)$, the price at time t of a stock derivative with payout $D(T,T)$ at T, as

$$D(t,T) = Z(t,T)E_*\left(D(T,T) \mid S_t\right),$$

where E_* is the risk-neutral expectation with respect to $Z(t, T)$. If the derivative has payout $D(T, T) = g(S_T)$ and the risk-neutral density is $f(S_T)$, then the option price is given by

$$D(t, T) = Z(t, T) \int g(x)f(x)dx.$$

This is an appealing result, and is often the point where derivatives professionals start their careers. On my first day in finance, I calculated an expected value then multiplied it by a discount factor. At the time it seemed a plausible approach to pricing derivatives, but in order to establish the result rigorously we needed the fundamental theorem.

In the past two chapters on replication, risk-neutrality and martingales, we have been exploring rich areas of derivatives pricing theory. We showed that options can be replicated by a portfolio of stock and bond, and that the replication price equals that given by discounted risk-neutral expectation. These two elements form the heart of quantitative finance. We showed the assumption of no-arbitrage is equivalent to the ratios of asset prices to the numeraire being martingales under the risk-neutral distribution. In particular, derivative prices can be calculated by taking discounted expected values of the derivative payout at maturity. We proved this for the binomial tree, and showed how we can change numeraire and the associated risk-neutral distribution.

In the next chapter we take the limit of the tree using the central limit theorem, and bridge to continuous time. It is often more elegant to work in continuous time where the manipulation of distributions and the calculation of the expected value of payouts is easier. We do not prove the fundamental theorem in continuous time. Instead we repeatedly appeal to the binomial tree proof and the intuition it provides, and assume that key results carry over to the continuous case. In particular, we do not concern ourselves with the mathematical conditions required to prove the fundamental theorem in continuous time. These conditions are almost always satisfied in the practical world, which is effectively discrete in time and value—for example, asset prices are only quoted to a finite number of decimal places.

9.6 EXERCISES

1. **Binomial tree: change of numeraire**
   Consider a one-step, two-state world where a stock has current price 100. After one year the stock is worth 110 with probability 0.8, and 90 with probability 0.2. One-year annually compounded interest rates are 5%.
   (a) Use the fundamental theorem to find the risk-neutral probability (of the stock being worth 110) with respect to the numeraires: (i) the money market account; (ii) the ZCB with maturity 1; and (iii) the stock.
   (b) Comment briefly on your answers to (a) (i) and (ii). In particular, can the risk-neutral probabilities with respect to the ZCB and money market account ever differ?

(c) By assuming no-arbitrage (thus $C_K(m,1)/N_m$ is a martingale for the appropriate numeraire and risk-neutral probability pair), price a one-year 105-strike call using the risk-neutral probabilities from (a)(i), (ii) and (iii). Verify the answers are the same.

2. **Binomial tree: random interest rates I**
Consider the two-step binomial tree in Chapter 8 Question 1. However, now suppose that if the stock is at 120, then the annually compounded interest rate from time $T = 1$ to $T = 2$ is 5%, not 10%.

(a) Write down the value of the money market account M_m at all states of the tree. Does the joint tree for (S_m, M_m) recombine?

(b) By using the martingale condition for S_m/M_m, find the risk-neutral probabilities with respect to the money market numeraire M_m at each node of the tree.

(c) Hence by using the martingale condition for $Z(m,2)/M_m$ show that

$$Z(0,2) = \left(\frac{65}{63}\right)\frac{1}{(1.1)^2}.$$

(d) Use (c) and an appropriate martingale condition to prove that the risk-neutral probability, with respect to the numeraire $Z(m,2)$, of the stock having value 120 at $T = 1$ is $44/65$. Hence show that the risk-neutral probabilities of this state, with respect to the money market account and the ZCB with maturity $T = 2$, differ by $2/195$. Do you want to revisit your comments in Question 1(b)?

3. **Binomial tree: random interest rates II**
A stock that pays no dividends has current price 100. In one year's time the stock price is 120 with probability 0.7, and 90 with probability 0.3. If the stock price at $T = 1$ is 120, then the price at $T = 2$ is 150 with probability 0.5, and 110 with probability 0.5. If the stock price at $T = 1$ is 90, then the price at $T = 2$ is 110 with probability 0.6, and 80 with probability 0.4. Annually compounded interest rates are 10%, except if the stock has price 120 at $T = 1$, in which case the interest rate from $T = 1$ to $T = 2$ is 20%.

(a) Draw the two-step binomial tree and give the value of the money market account at each state.

(b) Prove that the price $Z(0,2)$ at $T = 0$ of the ZCB with maturity $T = 2$ is

$$\left(\frac{1}{11}\right)\left(\frac{850}{99}\right).$$

(c) Find the price of a two-year 120-strike European call option.

(d) What is the risk-neutral distribution of the random variable $Z(1,2)$ with respect to the numeraire that is the ZCB with maturity $T = 2$?

4. **Binomial tree: FX forward brainteaser, part 2**
Let X_t be 'cable', that is, the price at t in dollars of one pound sterling. At $T = 0$ there are two dollars per pound (so $X_0 = \$2.00$). At $T = 1$, $X_1 = \$2.60$ (state A) with probability

0.5, and \$1.80 (state B) with probability 0.5. The one-year sterling interest rate $r_£$ and one-year dollar rate $r_\$$ are 2% and 4% respectively, both annually compounded. Assume the accrual factor $\alpha = 1$.

(a) Find Δ such that a portfolio of one \$2.00 call and Δ pounds sterling has the same value in both states at $T = 1$. Hence prove that the price at $T = 0$ of the \$2.00 call is

$$\left(\frac{3}{4}\right)\left(\frac{2}{1.02}\right) - \left(\frac{1.35}{1.04}\right) \quad (= \$0.17251).$$

Hint It is often helpful to think of '£1' as a stock (with, for example, price \$2.00 at $T = 0$). Remember this 'stock' pays interest, that is, your holding of it increases.

(b) Hence show that the risk-neutral probability (with respect to the dollar money market account) of state A is

$$p^* = \left(\left(\frac{3}{4}\right)\left(\frac{2}{1.02}\right) - \left(\frac{1.35}{1.04}\right)\right)\left(\frac{1.04}{0.60}\right) = 0.29902.$$

Verify that

$$E_{p^*}(X_1) = X_0\frac{(1 + r_\$)}{(1 + r_£)}.$$

(c) Restate the one-step model for the FX rate in terms of $Y_t = 1/X_t$, the value in pounds sterling of one dollar. By setting

$$E_{q^*}(Y_1) = Y_0\frac{(1 + r_£)}{(1 + r_\$)},$$

find q^*, the risk-neutral probability of A with respect to the pound sterling money market account.

(d) Use q^* to find by risk-neutral expectation the price of two £0.50 puts. Remember we are now working with a GBP asset so think of one dollar as a stock with price in pounds sterling. Is your answer the same as the price of the \$2.00 call from (a)?

5. **Call price as numeraire**

A stock that pays no dividends has current price S_0. In one year's time the stock price is 111 with probability 0.75 (state A), and 75 with probability 0.25 (state B). Annually compounded interest rates are 10%. The one-year 70-strike European call has price 10 at $T = 0$.

(a) Write down the payout of the 70-strike call at states A and B.

(b) By using a martingale condition for $C_{70}(m, 1)/M_m$, find the risk-neutral probability of state A with respect to the money market numeraire.

(c) Hence, or otherwise, prove that $S_0 = 810/11$ $(= 73.64)$.

(d) Find the price of a one-year 80-strike American put. For $K > 0$, write down a simple formula for the price of the one-year K-strike American put.

(e) By using an appropriate martingale condition, show that q^*, the risk-neutral probability of state A with respect to the 70-strike call numeraire, satisfies

$$q^* \left(\frac{11}{5} - \frac{11}{41} \right) = \frac{6}{5}.$$

(f) Which of the following assets cannot be used as a numeraire in this binomial tree?

I. The stock.

II. The one-year 100-strike European call.

III. A forward contract with delivery price 90 and maturity one year.

IV. The ZCB with maturity $T = 1$.

V. The one-year 115-strike European put.

6. Martingales and trading strategies

Your answer should comprise fewer lines than the statement of the question.

(a) A ***trading strategy*** can be thought of as holding Λ_{i+1} of asset X from time i to $i + 1$, where Λ_{i+1} is a sequence of random variables. Total profit at time $n + 1$ is thus given by

$$Y_{n+1} = \sum_{i=0}^{n} \Lambda_{i+1} \left(X_{i+1} - X_i \right).$$

Note Y is the discrete formulation of the ***stochastic integral*** $\int \Lambda dX$, central to modern probability theory. The lack of ability to see into the future is expressed mathematically by stating that Λ_{i+1} is ***previsible***, that is, it is a deterministic function of X_i known at time i. Show that if X_n is a martingale, then for a previsible trading strategy

$$E \left(Y_{n+1} - Y_n \mid X_n \right) = 0,$$

and hence Y_{n+1} is a martingale. In other words, you cannot beat the system.

(b) Suppose there are J traders, each with trading strategy $\Lambda_{i+1}^{(j)}, j = 1, \ldots, J$, not necessarily independent. Show that the total profit over all traders at time $n + 1$

$$W_{n+1} = \sum_{j=1}^{J} \sum_{i=0}^{n} \Lambda_{i+1}^{(j)} \left(X_{i+1} - X_i \right)$$

satisfies

$$E\left(W_{n+1} - W_n \mid X_n\right) = 0.$$

That is, a group cannot beat the system.

(c) Suppose a trading period lasts until time $n+1$ and that a trader is then compensated by the formula $\lambda \max\{Y_{n+1}, 0\}$, where $0 \leq \lambda \ll 1$ is the 'payout ratio' (for example, $\lambda = 5\%$). What strategy can two traders in collusion implement to ensure that between them they get paid, even if X is a martingale? How does the presence of a stop-loss affect this strategy?

7. **The normal model**
Show that if the risk-neutral distribution of S_T is given by $S_T \mid S_t \sim N\left(F, \psi^2(T-t)\right)$, where $F = F(t,T)$ is the forward price, then the price of a K-strike straddle is approximated by

$$Z(t,T)\sqrt{\frac{2}{\pi}}\psi\sqrt{T-t}\left(1+\frac{k^2}{2}\right), \text{ where } k = \frac{F-K}{\psi\sqrt{T-t}}.$$

Hint You may use your results from Question 1(b) and Question 4 in the preface.

10

Continuous-time limit and Black–Scholes formula

In this chapter we develop a framework that allows us to appeal to the central limit theorem and take the limit of the binomial tree as step sizes become smaller. We investigate two particular limiting cases which give rise to lognormal distributions for the stock price S_T, first under the real world probabilities and secondly in the risk-neutral case.

10.1 Lognormal limit

Suppose that the actual probability of an up or down move on the binomial tree is $1/2$. This assumption turns out to be unimportant to the results but makes the arithmetic neater. We define the ***logarithmic return***

$$\lambda_n = \log\left(\frac{S_n}{S_{n-1}}\right)$$

so $S_n = S_{n-1}e^{\lambda_n}$. Compare this to the ***multiplicative return*** l_n such that

$$l_n = \frac{S_n - S_{n-1}}{S_{n-1}}$$

and $S_n = S_{n-1}(1 + l_n)$. Then

$$\lambda_n = \begin{cases} \log(1+u) & \text{with probability } 1/2 \\ \log(1+d) & \text{with probability } 1/2. \end{cases}$$

Let S_T be the stock price at a fixed maturity T, and suppose there are N steps on the binomial tree of size ΔT, with $T = N\Delta T$. Define

$$Y_T = \log\left(\frac{S_T}{S_0}\right),$$

and let $E(Y_T) = \mu T$ and $\text{Var}(Y_T) = \sigma^2 T$ for some μ, σ. We will take limits under the constraint that the mean and variance of Y_T stay fixed as we increase the number of time steps N. In particular, u and d will depend on ΔT as we change the step size, although we will suppress this dependence in notation. Now,

$$Y_T = \log S_T - \log S_0 = \log\left(\frac{S_{N\Delta T}}{S_0}\right) = \log\left(\frac{S_{N\Delta T}}{S_{(N-1)\Delta T}}\right) + \ldots + \log\left(\frac{S_{\Delta T}}{S_0}\right)$$
$$= \lambda_N + \lambda_{N-1} + \ldots + \lambda_1,$$

where the λ_i are independent and identically distributed (IID) with

$$E(\lambda_i) = \mu\frac{T}{N} = \mu\Delta T \text{ and } \text{Var}(\lambda_i) = \sigma^2\Delta T.$$

Given λ_i is a random variable that can take only two possible values, we must have

$$\lambda_i = \begin{cases} \mu\Delta T + \sigma\sqrt{\Delta T} & \text{with probability } 1/2 \\ \mu\Delta T - \sigma\sqrt{\Delta T} & \text{with probability } 1/2. \end{cases}$$

That is, we have $\log(1+u) = \mu\Delta T + \sigma\sqrt{\Delta T}$ and $\log(1+d) = \mu\Delta T - \sigma\sqrt{\Delta T}$. So we can write

$$\log S_{N\Delta T} = \log S_{(N-1)\Delta T} + \mu\Delta T + \sigma\sqrt{\Delta T}\,\xi,$$

where $\xi = \pm 1$ with probability $1/2$. Therefore,

$$\log S_T = \log S_0 + \mu T + \sigma\sqrt{T}\frac{1}{\sqrt{N}}\sum_{i=1}^{N}\xi_i,$$

where the ξ_i are IID ± 1 with probability $1/2$. Letting $N \to \infty$, the central limit theorem gives

$$\log S_T = \log S_0 + \mu T + \sigma\sqrt{T}\,W, \text{ where } W \sim N(0,1).$$

Therefore, S_T is lognormally distributed in this limiting case. The term σ is called the ***volatility*** of the stock. We can alternatively write

$$S_T = S_0\, e^{\mu T + \sigma\sqrt{T}W}$$

and we have (see Question 2(a) in the preface)

$$E(S_T|S_0) = S_0 e^{\mu T + \frac{1}{2}\sigma^2 T}.$$

10.2 Risk-neutral limit

On the binomial tree with time steps of length ΔT, the risk-neutral probability is given by

$$p^* = \frac{r\Delta T - d}{u - d},$$

where r is the constant interest rate for the term of the step. On this tree

$$\log(1 + u) = \mu\Delta T + \sigma\sqrt{\Delta T},$$

and so for small ΔT

$$1 + u = e^{\mu\Delta T + \sigma\sqrt{\Delta T}} = 1 + \mu\Delta T + \sigma\sqrt{\Delta T} + \frac{\sigma^2}{2}\Delta T + O\left(\Delta T^{\frac{3}{2}}\right).$$

Similarly, we have

$$1 + d = 1 + \mu\Delta T - \sigma\sqrt{\Delta T} + \frac{\sigma^2}{2}\Delta T + O\left(\Delta T^{\frac{3}{2}}\right).$$

Therefore,

$$\frac{r\Delta T - d}{u - d} = \frac{r\Delta T - \mu\Delta T + \sigma\sqrt{\Delta T} - \frac{\sigma^2}{2}\Delta T + O\left(\Delta T^{\frac{3}{2}}\right)}{2\sigma\sqrt{\Delta T} + O\left(\Delta T^{\frac{3}{2}}\right)},$$

and we obtain

$$p^* = \frac{1}{2} + \frac{1}{2}\left(\frac{r - \mu - \frac{\sigma^2}{2}}{\sigma}\right)\sqrt{\Delta T} + O(\Delta T).$$

We note the second term on the right hand side is an adjustment term to obtain the risk-neutral probability p^* from the real world probability $1/2$ of an up move. Our assumption that the latter is $1/2$ simplifies the arithmetic, but is unimportant. The adjustment to obtain the risk-neutral probabilities would change accordingly.

So under the risk-neutral probabilities,

$$\log S_T = \log S_0 + \sum_{i=1}^{N} \lambda_i = \log S_0 + \sum_{i=1}^{N} (\mu\Delta T + \sigma\sqrt{\Delta T}\,\xi_i^*),$$

where the ξ_i^* are IID,

$$\xi_i^* = \begin{cases} +1 & \text{with probability } p^* \\ -1 & \text{with probability } 1 - p^*. \end{cases}$$

Since $N\Delta T = T$, we rearrange to obtain

$$\log S_T = \log S_0 + \mu T + \sigma\sqrt{T}\sqrt{\frac{1}{N}}\sum_{i=1}^{N}\xi_i^*.$$

The mean and variance of ξ_i^* are given by

$$E(\xi_i^*) = \nu = \left(\frac{r - \mu - \frac{1}{2}\sigma^2}{\sigma}\right)\sqrt{\Delta T}$$

$$\text{and } \operatorname{Var}(\xi_i^*) = E(\xi_i^{*2}) - (E(\xi_i^*))^2 = 1 - \nu^2 = 1 - \left(\frac{r - \mu - \frac{\sigma^2}{2}}{\sigma}\right)^2 \Delta T.$$

We now use the central limit theorem to give

$$\frac{1}{\sqrt{N}}\sum_{i=1}^{N}\xi_i^* \sim N\left(\sqrt{N}\sqrt{\Delta T}\left(\frac{r - \mu - \frac{1}{2}\sigma^2}{\sigma}\right), 1\right) \text{ as } N \to \infty.$$

Equivalently, as $N \to \infty$,

$$\frac{1}{\sqrt{N}}\sum_{i=1}^{N}\xi_i^* = \sqrt{T}\left(\frac{r - \mu - \frac{1}{2}\sigma^2}{\sigma}\right) + W, \text{ where } W \sim N(0, 1).$$

Substituting, we obtain

$$\log S_T = \log S_0 + \left(r - \frac{1}{2}\sigma^2\right)T + \sigma\sqrt{T}W, \text{ where } W \sim N(0, 1).$$

That is, under the risk-neutral probability, S_T is lognormally distributed with

$$E_*(\log S_T \mid S_0) = \log S_0 + \left(r - \frac{1}{2}\sigma^2\right)T \text{ and } \operatorname{Var}_*(\log S_T \mid S_0) = \sigma^2 T.$$

The actual expectation μ does not appear in the risk-neutral distribution.

Note that the risk-neutral expectation for $\log S_T$ differs from the expectation under the actual probabilities, but its volatility is the same. Under the risk-neutral lognormal limit,

$$E_*(S_T \mid S_0) = S_0 e^{rT}.$$

The choice of current time t (above set to be zero) is unimportant and we have the general result that

$$\log S_T \mid S_t \sim N\left(\log S_t + \left(r - \frac{1}{2}\sigma^2\right)(T-t), \sigma^2(T-t)\right).$$

10.3 Black–Scholes formula

The fundamental theorem states that

$$C_K(t,T) = Z(t,T)E_*\left((S_T - K)^+ \mid S_t\right).$$

Furthermore, under the risk-neutral lognormal limit above we found that

$$\log S_T \mid S_t \sim N\left(\log S_t + \left(r - \frac{1}{2}\sigma^2\right)(T-t), \sigma^2(T-t)\right).$$

Rather than work with the lognormal density directly, it is easier to express the option payout in terms of $Y_T = \log S_T$, and use the normal density function to calculate the expected value. We can write

$$\begin{aligned} C_K(t,T) &= Z(t,T)E_*\left((e^{Y_T} - K)^+ \mid S_t\right) \\ &= Z(t,T)\int_{\log K}^{\infty} (e^y - K)\frac{1}{\sqrt{2\pi}\sigma\sqrt{T-t}} e^{-\frac{(y-\nu)^2}{2\sigma^2(T-t)}}\, dy, \end{aligned}$$

where

$$\nu = \log S_t + \left(r - \frac{1}{2}\sigma^2\right)(T-t).$$

Note that the limits of integration on Y_T run from $\log K$ to ∞.

Note This is the most important tail integral in finance, and I believe everyone should do it at least once. Revisiting it every couple of years after starting work on Wall Street is also a good way of keeping technically fresh.

Result *We obtain (see exercises)*

$$C_K(t,T) = S_t\Phi(d_1) - KZ(t,T)\Phi(d_2),$$

where

$$d_1 = \frac{\log\left(\frac{S_t}{K}\right) + \left(r + \frac{1}{2}\sigma^2\right)(T-t)}{\sigma\sqrt{T-t}} \quad \textit{and } d_2 = d_1 - \sigma\sqrt{T-t},$$

the Black–Scholes formula.

We can rewrite the Black–Scholes formula in the form

$$C_K(t,T) = Z(t,T)\,(F(t,T)\Phi(d_1) - K\Phi(d_2)),$$

where both $F(t, T)$ and K are 'time T' quantities, amounts that are exchanged at T, whereas S_t is a 'time t' quantity.

Note Observe how the Black–Scholes formula, a continuous-time result, has a similar form to the discrete-time Cox Ross Rubinstein formula obtained in Chapter 8.

The formula is named after Fischer Black and Myron Scholes who, along with Robert Merton, were pioneers of continuous-time finance in the 1970s. Black died in 1995—there was a sense of sadness and respect among options traders that day—but Scholes and Merton went on to win the Nobel prize in economic sciences in 1997 for their work on option pricing.

We obtained the Black–Scholes formula by calculating an expected value, drawing on the fundamental theorem of asset pricing. This route via martingales and expected values is particularly appealing to probabilists. The original derivation in Black and Scholes (1973) is significantly different, and we outline alternative approaches to deriving the formula using stochastic processes, Ito calculus and partial differential equations in Chapter 16.

The Black–Scholes formula is one element of quantitative finance that has penetrated broader public awareness, featuring in general interest news as derivatives received heightened scrutiny during the financial crisis. In our exposition the formula has a straightforward interpretation as the expected value of the option payout under the lognormal risk-neutral distribution. The formula enables one to translate between a particular underlying dynamic for the asset—the limit of the binomial tree—and a unique arbitrage-free option price. Whilst plausible, there is of course no reason why in practice the risk-neutral distribution should be lognormal. In Chapter 11 we explore the key duality between option prices and probability distributions, and show that the Black–Scholes formula can still be a useful tool even when lognormality does not hold.

10.4 Properties of Black–Scholes formula

The Black–Scholes formula is easily coded, and a good understanding of its properties comes from graphs of its value as a function of S_t, for different volatilities. Figure 10.1 shows the price of a one-year call when $r = 0$, for $\sigma = 0.1, 0.2, 0.3, 0.4$ and 0.5.

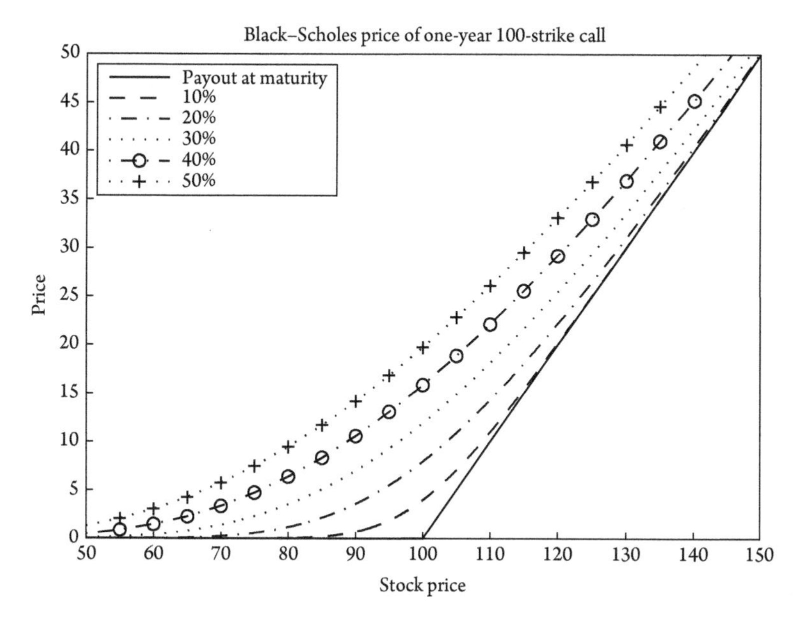

Figure 10.1 Black–Scholes call price

We can analytically investigate several limiting cases.

Result *As* $S_t \to \infty$, $\Phi(d_1) \to 1$ *and* $\Phi(d_2) \to 1$, *since* $\log\left(\frac{S_t}{K}\right) \to \infty$. *Therefore, as* $S_t \to \infty$,

$$C_K(t,T) \to S_t - Ke^{-r(T-t)},$$

the value of a forward contract.

This makes sense since as $S_t \to \infty$, we are certain to exercise the call option, and thus the option price tends to the value of being long a forward contract.

Result *As* $\sigma \to 0$,

$$d_1 = \frac{\log\left(\frac{S_t}{K}\right) + (r + \frac{1}{2}\sigma^2)(T-t)}{\sigma\sqrt{T-t}} \to \begin{cases} +\infty & \text{if } \log(S_t/K) + r(T-t) > 0 \\ -\infty & \text{if } \log(S_t/K) + r(T-t) < 0. \end{cases}$$

The same result holds for d_2. *Therefore,*

$$C_K(t,T) \to \begin{cases} S_t - Ke^{-r(T-t)} & \text{if } S_t > Ke^{-r(T-t)} \\ 0 & \text{if } S_t \leq Ke^{-r(T-t)}. \end{cases}$$

As the volatility of the stock tends to zero, the stock price becomes deterministic, equivalent to a holding of cash invested at rate r. If the stock is in-the-money-forward, then the call option becomes equivalent to a long forward contract. If it is out-of-the-money-forward, the call option is worthless.

Result *As $\sigma \to \infty$, $\Phi(d_1) \to 1$ and $\Phi(d_2) \to 0$, and thus $C_K(t,T) \to S_t$.*

Recall the bounds on a European call that we obtained in Chapter 7, namely that

$$\max\{0, S_t - KZ(t,T)\} \leq C_K(t,T) \leq S_t.$$

Our results show that these upper and lower bounds are in fact tight and cannot be improved, since we have presented cases where the call price becomes arbitrarily close to the bounds.

We can also derive a good approximation for the at-the-money-forward straddle price under Black–Scholes. Suppose

$$K = F(t,T) \left(= S_t e^{r(T-t)} \right),$$

then

$$C_K(t,T) = P_K(t,T) = S_t \Phi\left(+\frac{1}{2}\sigma\sqrt{T-t} \right) - S_t \Phi\left(-\frac{1}{2}\sigma\sqrt{T-t} \right).$$

Result *Using a Taylor series expansion for the normal cumulative distribution function $\Phi(\cdot)$, we find that the straddle price is approximated by*

$$S_t \sigma\sqrt{T-t}\sqrt{\frac{2}{\pi}}\left(1 - \frac{\sigma^2(T-t)}{24}\right) + \text{higher order terms}.$$

Comparing to the straddle price under the normal distribution (Chapter 9)

$$Z(t,T)\,\psi\sqrt{T-t}\sqrt{\frac{2}{\pi}},$$

we find that there is a correspondence,

$$\psi \longleftrightarrow \sigma F(t,T)\left(1 - \frac{\sigma^2(T-t)}{24}\right), \text{ or approximately } \psi \longleftrightarrow \sigma F(t,T),$$

obtained by equating straddle prices. This is one pragmatic method practitioners use to switch between approximately equivalent normal and lognormal distributions, both of which are often utilized in practice as reasonable candidates for the risk-neutral distribution.

As an example, suppose for a particular asset, the forward price $F(t,T) = \$50$. A lognormal volatility of 15% corresponds approximately to a normal volatility (standard deviation) of \$7.5 per annum.

10.5 Delta and vega

The ***delta*** of an option is the partial derivative of its price with respect to the underlying asset price. Including explicitly here the dependence of the call price on the stock price S_t, for small ΔS_t we have

$$C_K(t,T,S_t+\Delta S_t) \approx C_K(t,T,S_t) + \Delta S_t \frac{\partial C_K(t,T,S_t)}{\partial S_t}.$$

The delta measures how much the option price changes for small changes in the stock price.

Under the Black–Scholes formula,

$$\frac{\partial C_K(t,T,S_t)}{\partial S_t} = \Phi(d_1) + S_t \frac{\partial}{\partial S_t}\Phi(d_1) - Ke^{-r(T-t)}\frac{\partial}{\partial S_t}\Phi\left(d_1 - \sigma\sqrt{T-t}\right).$$

It is probably worth the discipline of doing this differentiation once. The last two terms involve use of the chain rule, although there is a simplifying trick, and eventually cancel to give

$$\frac{\partial C_K(t,T,S_t)}{\partial S_t} = \Phi(d_1).$$

We immediately conclude

$$0 \le \frac{\partial C_K(t,T,S_t)}{\partial S_t} \le 1.$$

The delta of a call under the Black–Scholes formula is shown in Figure 10.2. If the call is ATMF with $K = S_t e^{r(T-t)}$, then its delta equals $\Phi(\sigma\sqrt{T-t}/2)$. The delta of an ATMF call under Black–Scholes is close to, but not equal to, $1/2$.

Note that a portfolio consisting of long one call and short an amount $\frac{\partial C_K(t,T,S_t)}{\partial S_t}$ of stock is instantaneously hedged to movements in the stock price (compare to the binomial tree delta hedge), and thus over a small time interval is a replicating portfolio for the money market account, growing instantaneously at rate r. This is the basis for deriving the Black–Scholes formula via partial differential equations, which we sketch briefly in Chapter 16.

The ***vega*** (sometimes called kappa by those wishing to remain within the Greek alphabet) of an option is the partial derivative of its price with respect to volatility σ, and measures the exposure of an option or other derivative contract price to movements in volatility.

For a call option under the Black–Scholes formula, its vega is

$$\frac{\partial C_K(t,T)}{\partial \sigma} = S_t\sqrt{T-t}\,\phi(d_1) > 0,$$

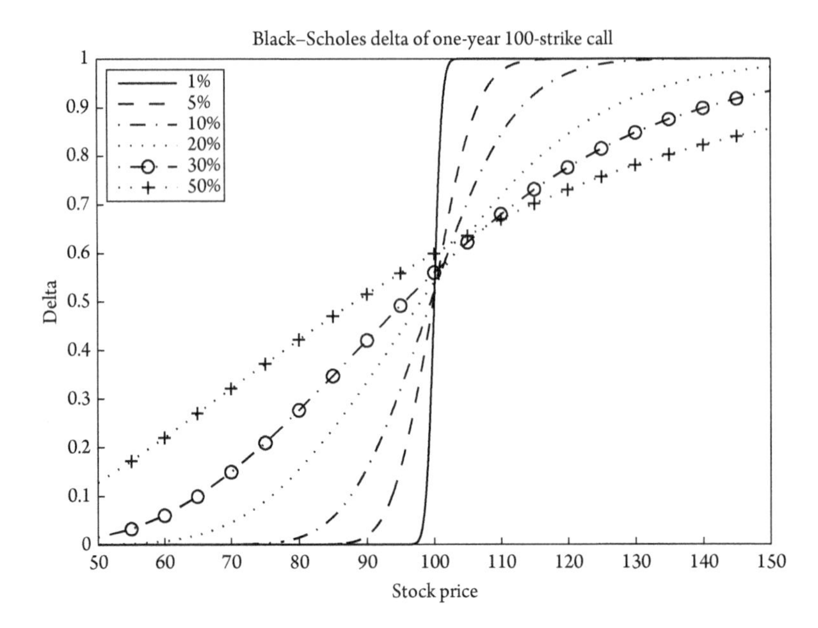

Figure 10.2 Black–Scholes delta

where $\phi(\cdot)$ is the standard normal probability density function. Using a Taylor series expansion we find that the vega of an ATMF straddle equals

$$S_t\sqrt{\frac{2}{\pi}}\left(1 - \frac{\sigma^2}{8}\right) + \text{higher order terms.}$$

The vega of a call under the Black–Scholes formula is shown in Figure 10.3.

The vega of a call, put or a straddle is positive. Such a position is often termed as being 'long vega'. The holder of a straddle, for example, would want volatility to go up. We introduce, therefore, the concept of trading volatility, buying a straddle if one believes volatility is likely to increase.

The vega of a forward contract, swap and FRA are all zero, since their value can be determined simply from the current prices of the stock or ZCBs. However, as we noted in Chapter 5, the vega of a future is not zero. It will be positive if the underlying asset is positively correlated with interest rates, since the convexity correction is proportional to the covariance. A long Eurodollar position, for example, is a short vega position.

We know the vega of a long call position is positive, and that of a short call is negative, but what about the vega of a call spread, digital or butterfly? Intuitively, a contract is long vega if one would profit if volatility were to increase, and the stock be more likely to move more. For instance, the vega of a K_1, K_2 call spread is positive when $S_t \ll K_1$, and negative when $S_t \gg K_2$. The precise point at which the vega is zero will depend on the option pricing model.

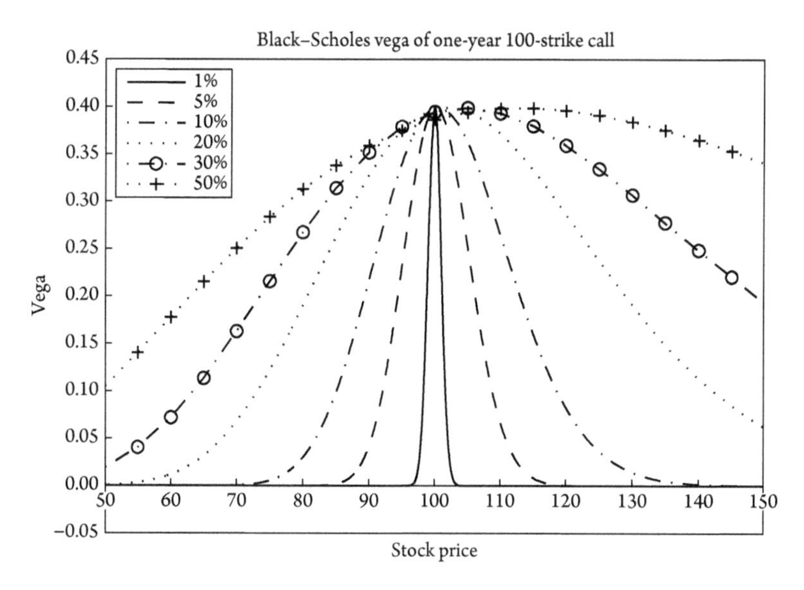

Figure 10.3 Black–Scholes vega

Similarly, the vega of a call butterfly changes from positive to negative to positive as S_t increases from below the lowest strike to above the highest.

One can calculate the vega analytically under the Black–Scholes formula for a range of option structures, simply by differentiating the price function with respect to σ. However, good intuition regarding the vega of call spreads, butterflies and other option structures can be obtained by inspecting the price graphs as a function of S_t, for different values of σ. Figure 10.4 shows the price of a one-year 100, 110 call spread and Figure 10.5 the price of a one-year 90, 100, 110 call butterfly, for $\sigma = 0.05, 0.1, 0.15, 0.2$ and 0.5. By observing how the price changes for different σ we note, for example, that the vega of the butterfly at $S_t = 100$ is negative and at $S_t = 65$ is positive for the values of σ considered. However, around $S_t = 85$ the price of the butterfly first increases as volatility increases, then decreases. In other words, the vega at $S_t = 85$ can be either positive or negative depending on the level of σ. Question 7 in the exercises explores the vega profiles of several option structures.

Recall that put-call parity states

$$C_K(t,T) - P_K(t,T) = V_K(t,T) = S_t - KZ(t,T).$$

Differentiating with respect to S_t, we obtain

$$\frac{\partial C_K(t,T)}{\partial S_t} - \frac{\partial P_K(t,T)}{\partial S_t} = 1.$$

Given that the delta of the ATMF call is not necessarily $1/2$, the delta of the ATMF put and call will in general differ in magnitude.

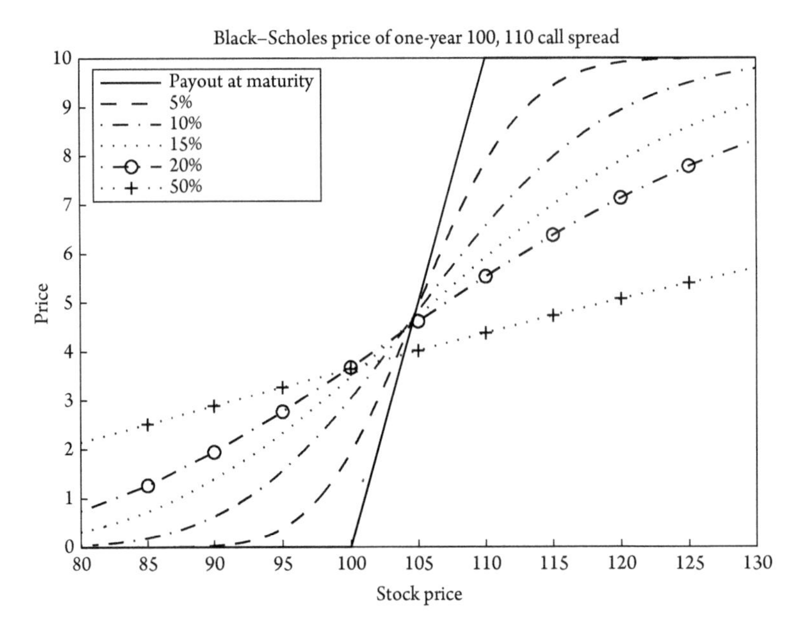

Figure 10.4 Black–Scholes price of call spread

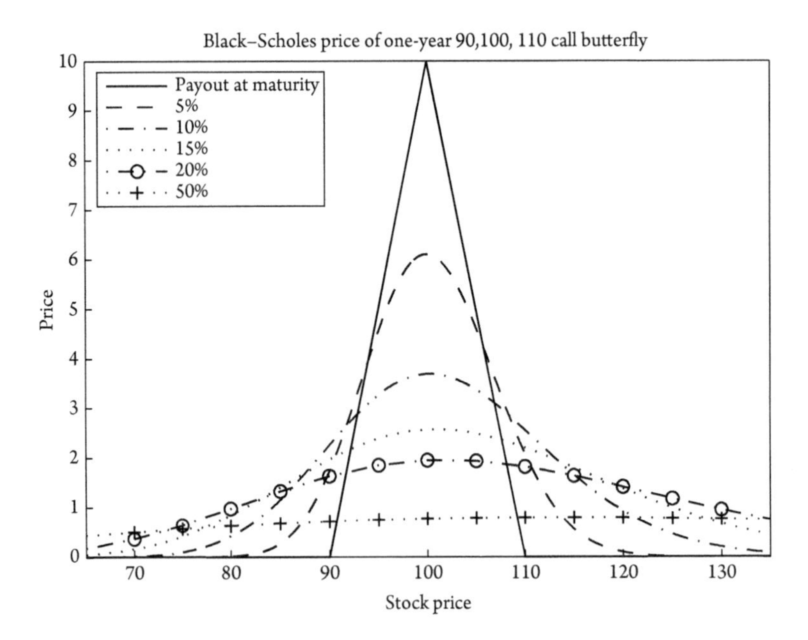

Figure 10.5 Black–Scholes price of call butterfly

Differentiating with respect to σ we obtain

$$\frac{\partial C_K(t,T)}{\partial \sigma} - \frac{\partial P_K(t,T)}{\partial \sigma} = 0.$$

Therefore, the vega of a put always equals the vega of a call of the same strike and maturity.

For the ATMF strike $K = F(t,T)$, we have $C_{F(t,T)}(t,T) = P_{F(t,T)}(t,T)$. However, note that

$$\frac{\partial C_{F(t,T)}(t,T)}{\partial S_t} \neq \frac{\partial P_{F(t,T)}(t,T)}{\partial S_t}$$

since $F(t,T)$ itself is a function of S_t.

10.6 Incorporating random interest rates

We saw we could rewrite the Black–Scholes formula in the form

$$C_K(t,T) = Z(t,T)\left(F(t,T)\Phi(d_1) - K\Phi(d_2)\right)$$

$$\text{where } d_1 = \frac{\log\left(\frac{F(t,T)}{K}\right) + \frac{1}{2}\sigma^2(T-t)}{\sigma\sqrt{(T-t)}} \text{ and } F(t,T) = \frac{S_t}{Z(t,T)}.$$

This suggests an approach to pricing stock options when interest rates are random variables and we can no longer assume that $Z(t,T) = e^{r(T-t)}$ for a constant r and for all $t \leq T$. First, note that by the fundamental theorem $F(t,T) = S_t/Z(t,T)$ is a martingale under the risk-neutral distribution with respect to the $Z(t,T)$ numeraire. Secondly, in Chapter 7 we showed that the price of a T-expiry K-strike call on the stock equals the price of a call on a forward contract with delivery price K and same maturity T, since $F(T,T) = S_T$. Therefore, if we work solely with the forward price, and assume that the risk-neutral distribution for $F(t,T)$ is given by

$$F(T,T) \sim \text{lognormal}\left(\log(F(t,T)) - \frac{1}{2}\sigma^2(T-t), \sigma^2(T-t)\right),$$

then we obtain the Black–Scholes equation (again) while incorporating random interest rates. The key difference is that now σ represents the volatility of the forward price, not the stock price. That is, we capture the joint random process of S_t and $Z(t,T)$ in the process for $F(t,T)$. Working with the stock and interest rate (or ZCB) separately, when both are random, can often be challenging.

10.7 EXERCISES

1. **The Black–Scholes formula**

   (a) Suppose for $t \leq T$, a stock that pays no dividends has risk-neutral distribution $S_T \mid S_t$ given by

   $$\log S_T|S_t \sim N\left(\nu, \sigma^2(T-t)\right), \text{ where } \nu = \log S_t + \left(r - \frac{1}{2}\sigma^2\right)(T-t)$$

   and r is the continuous interest rate. Prove that the price at time t of a K-strike call with exercise date T is given by

   $$C_K(t,T) = Z(t,T)\left(F(t,T)\Phi\left(d_1\right) - K\Phi\left(d_2\right)\right),$$

   where

   $$d_1 = \frac{\log\left(\frac{S_t}{K}\right) + \left(r + \frac{1}{2}\sigma^2\right)(T-t)}{\sigma\sqrt{T-t}} \text{ and } d_2 = d_1 - \sigma\sqrt{T-t}.$$

   Hint (i) Let $S_T = e^{Y_T}$ where Y_T is normally distributed. (ii) Be careful about the range of integration for Y. (iii) Use the identity

   $$y - \frac{(y-\nu)^2}{2\sigma^2\tau} = -\frac{\left(y-(\nu+\sigma^2\tau)\right)^2}{2\sigma^2\tau} + \left(\nu + \frac{1}{2}\sigma^2\tau\right), \text{ where } \tau = (T-t).$$

   (b) Use put-call parity to prove that the price $P_K(t,T)$ of a European put is given by

   $$P_K(t,T) = Z(t,T)(K\Phi\left(-d_2\right) - F(t,T)\Phi\left(-d_1\right)).$$

2. **A knockout option**

   (a) Use your results from Question 1 to prove that, for $\lambda > 0$, one call with strike $\lambda F(t,T)$ has the same value as λ puts with strike $F(t,T)/\lambda$.

   (b) Suppose that $K_1 > K_2$ and, at current time t, $F(t,T) > K_2$. Use (a) to prove that a K_1-strike call with maturity T knocking out if the forward price hits K_2 at any time before T has price at time t equal to

   $$C_{K_1}(t,T) - \frac{K_1}{K_2}P_{K^*}(t,T),$$

   for a particular strike K^*. Find the expression for K^* in terms of K_1 and K_2.

   Hint Find a linear combination of a call and a put that has zero value at T^* when $F(T^*,T) = K_2$, where $t < T^* < T$.

3. **Power options I**
A K-strike ***power call*** is an option with payout $(S_T - K)^2$ at time T if $S_T > K$, and zero otherwise.

Use the fundamental theorem to calculate the price of a power call under the Black–Scholes model. You may use any result from Question 1.

4. **Power options II**
A ***capped power call*** is a K_1-strike power call whose payout is capped at $K_2 > K_1$, that is, it has payout at time T

$$\begin{cases} 0 & \text{if } S_T < K_1 \\ (S_T - K_1)^2 & \text{if } K_1 \leq S_T \leq K_2 \\ (K_2 - K_1)^2 & \text{if } S_T > K_2. \end{cases}$$

Capped power calls were briefly popular in derivative markets in the mid-1990s.

(a) Write down (but do not solve) the integral expression for the price of the capped power call.

(b) Suppose $K_1 = 100$ and $K_2 = 115$, and that you observe call option prices for $K = 100, 105, 110$ and 115. By comparing the payout functions of the capped power call with linear combinations of calls, find upper and lower bounds for the price of the capped power call in terms of the four call prices. Is there a least expensive upper bound and a most expensive lower bound? If so, find them.

5. **The squared payout**
A derivative contract has payout at maturity defined by $D(T, T) = S_T^2$, and price $D(t, T)$ at current time t.

(a) One trader states that the replicating portfolio for this derivative is a holding of S_t stocks, and thus $D(t, T) = S_t^2$. What is wrong with this argument?

(b) Another trader states that the replicating portfolio is a holding of S_T stocks, and thus $D(t, T) = S_t S_T$. What is wrong with this argument?

(c) Suppose the risk-neutral distribution of S_T conditional on S_t, with respect to the money market numeraire, is lognormal $(\log S_t + \nu(T - t), \sigma^2(T - t))$, that is, $\log S_T \mid S_t \sim N(\log S_t + \nu(T - t), \sigma^2(T - t))$. Use a martingale condition for S_T/M_T to find ν in terms of σ and r, the constant, continuously compounded interest rate.

(d) Use the fundamental theorem to write down, but do not solve, an integral expression for $D(t, T)$ involving a normal probability density function.

(e) Suppose you are also told that the risk-neutral distribution of S_T conditional on S_t, with respect to the stock numeraire, is lognormal $(\log S_t + \gamma(T - t), \sigma^2(T - t))$. Use an appropriate martingale condition, and the fact that $1/S_T$ is also lognormal, to find γ in terms of r and σ.

(f) Hence use the fundamental theorem to prove that, assuming the distribution in (e) holds,

$$D(t,T) = S_t^2 e^{(r+\sigma^2)(T-t)}.$$

6. **Normal versus lognormal distributions**
Market participants often convert between normal volatility ψ and lognormal volatility σ by adopting the approximate equivalence $\psi \longleftrightarrow \sigma F(t,T)$. We will investigate two others.
   (a) Setting $K = F(t,T) = S_t e^{r(T-t)}$, and using the Taylor series for $\Phi(\cdot)$, show that the Black–Scholes price at t of an at-the-money-forward straddle is approximated by

$$S_t \sigma \sqrt{T-t} \sqrt{\frac{2}{\pi}} \left(1 - \frac{\sigma^2 (T-t)}{24}\right).$$

   By equating this to the price of an ATMF straddle under the normal model with standard deviation ψ, find an equivalence between σ and ψ.
   (b) Calculate $\mathrm{Var}(S_T)$ under the Black–Scholes model (Question 2(a) in the preface may be useful), and hence find an equivalence between ψ and σ by equating $\psi^2(T-t)$, which is $\mathrm{Var}(S_T)$ under the normal model, to your answer.

7. **Vega and delta**
For each of the following positions I-X in Table 10.1, determine whether Bank A has vega ≥ 0, Bank A has vega ≤ 0, there is no volatility exposure, or the volatility exposure is indeterminate ('?'). Similarly, determine Bank A's delta with respect to the stock price.

Table 10.1 Vega and delta of option positions.

		(i) Vega	**(ii) Delta**
	Bank A position	**$\geq 0, \leq 0, = 0$ or ?**	**$\geq 0, \leq 0, = 0$ or ?**
I	long a K_1 call		
II	short a K_1 put		
III	short a K_2, K_1 put spread		
IV	long a $K_1, (K_1+K_2)/2, K_2$ call butterfly		
V	long a K_1 put knocking out if $S_T < K_2$		
VI	long a K_2 put knocking out if $S_T > K_1$		
VII	short a K_1 call and long a K_1 put		
VIII	long a K_1 straddle		
IX	short a forward contract on the stock		
X	short a futures contract on the stock		

All options are European with maturity T on a stock that pays no dividends, $0 < K_1 < K_2$.

8. **The Black–Scholes formula by change of numeraire**
   Let S_t be the price at time t of a stock that pays no dividends.
   (a) Draw the payout functions for the two options
      (i) $KI\{S_T \geq K\}$
      (ii) $S_T I\{S_T \geq K\}$
   where $I\{\}$ is the indicator function. Which linear combination of (i) and (ii) gives the payout of a call option?
   (b) Suppose the risk-neutral distributions of S_T conditional on S_t with respect to the money market and stock numeraires are given by the answers to Question 5(c) and (e), respectively. Price options (i) and (ii) in terms of the normal cumulative distribution function $\Phi(\cdot)$ using the money market numeraire for (i) and the stock numeraire for (ii). Hence derive immediately the Black–Scholes equation for the price of a call option.

11

Option price and probability duality

In Chapters 2 and 3 we were able to determine the forward price and forward libor rates without any assumptions about the distribution of the random stock price or libor rate. In Chapter 7 we saw that the call option price did depend on the possible future states for the stock and hence its risk-neutral distribution. Here we derive the result that option prices in fact determine entirely the risk-neutral distribution of S_T.

11.1 Digitals and cumulative distribution function

For $\lambda > 0$, consider a portfolio consisting of λ $(K, K + 1/\lambda)$ call spreads, that is, long an amount λ of K calls and short λ $(K + 1/\lambda)$ calls. This portfolio has value at time t equal to

$$\lambda \left(C_K(t,T) - C_{K+\frac{1}{\lambda}}(t,T) \right),$$

and has payout at time T shown in Figure 11.1, equal to

$$\begin{cases} 0 & \text{if } S_T < K \\ \lambda(S_T - K) & \text{if } K \le S_T \le K + \frac{1}{\lambda} \\ 1 & \text{if } S_T > K + \frac{1}{\lambda}. \end{cases}$$

As $\lambda \to \infty$, the payout of the call spread tends to the payout of a digital call option struck at K, and we have established the following result.

Result *The price at time t of the digital call option struck at K equals*

$$\lim_{\lambda \to \infty} \lambda \left(C_K(t,T) - C_{K+\frac{1}{\lambda}}(t,T) \right) = -\frac{\partial C_K(t,T)}{\partial K}.$$

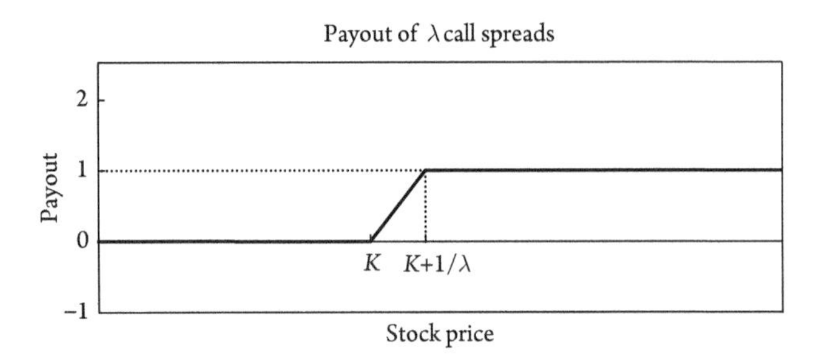

Figure 11.1 Payout of λ call spreads

We know this price must be positive, either by using the monotonicity theorem applied to the digital call payout, or from our elementary properties of call prices. We can also price the digital option directly via the fundamental theorem.

Result *Defining the payout at T of the digital call option to be*

$$D_K(T,T) = I\{S_T > K\},$$

then by the fundamental theorem, $D_K(t,T)$, its price at time t, satisfies

$$\frac{D_K(t,T)}{Z(t,T)} = E_*\left(\frac{D_K(T,T)}{Z(T,T)}\right) = E_*(I\{S_T > K\}) = P^*(S_T > K \mid S_t).$$

The price of the digital is thus simply the present value of the probability of receiving a payout, namely the stock price being above K at T.

Combining our results, we obtain

$$P^*\left(S_T \leq K \mid S_t\right) = 1 + \frac{1}{Z(t,T)}\frac{\partial C_K(t,T)}{\partial K}.$$

We have thus recovered the cumulative distribution function of the risk-neutral distribution of S_T from call prices.

We can differentiate again and obtain the risk-neutral density for $S_T \mid S_t$ given by

$$f_{S_T|S_t}(x) = \frac{1}{Z(t,T)}\left.\frac{\partial^2 C_K(t,T)}{\partial K^2}\right|_x,$$

where the right hand side is evaluated at $K = x$. This has to be positive (as we require for a density function) due to the convexity of call prices.

For the Black–Scholes model

$$P(S_T > K \mid S_t) = P(\log S_T > \log K \mid S_t)$$
$$= 1 - \Phi\left(\frac{\log K - \log S_t - \left(r - \frac{1}{2}\sigma^2\right)(T-t)}{\sigma\sqrt{T-t}}\right) = \Phi(d_2).$$

Thus we obtain the result

$$-\frac{\partial C_K(t,T)}{\partial K} = Z(t,T)\Phi(d_2),$$

simply by comparing the payout at maturity of the limiting call spread with the digital call, and without differentiation. Note that the price of the digital call option under the Black–Scholes formula is not equal to 1/2 when $K = S_t e^{r(T-t)}$, since for the lognormal distribution its mean $S_t e^{r(T-t)}$ does not equal its median.

11.2 Butterflies and risk-neutral density

For $\lambda > 0$, consider the call butterfly defined as the portfolio consisting of

$$\begin{cases} \lambda \text{ calls} & \text{with strike } K - \frac{1}{\lambda} \\ -2\lambda \text{ calls} & \text{with strike } K \\ \lambda \text{ calls} & \text{with strike } K + \frac{1}{\lambda}. \end{cases}$$

This butterfly has payout at maturity T shown in Figure 11.2, and price at t given by

$$\lambda\left(C_{K-\frac{1}{\lambda}}(t,T) - C_K(t,T)\right) - \lambda\left(C_K(t,T) - C_{K+\frac{1}{\lambda}}(t,T)\right),$$

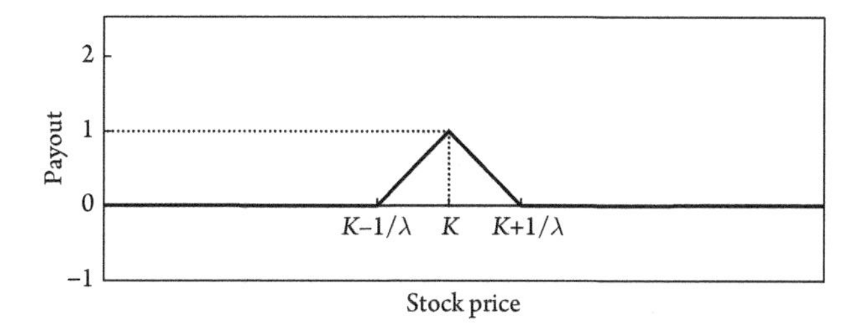

Figure 11.2 Payout of λ call butterflies

which we denote $B_{K,\lambda}(t, T)$. For large λ we can write

$$\lambda B_{K,\lambda}(t, T) \approx \frac{\frac{\partial C_K(t,T)}{\partial K}|_K - \frac{\partial C_K(t,T)}{\partial K}|_{K-\frac{1}{\lambda}}}{\frac{1}{\lambda}} \rightarrow \frac{\partial^2 C_K(t, T)}{\partial K^2} \text{ as } \lambda \rightarrow \infty.$$

We also know that

$$\left.\frac{\partial^2 C_K(t, T)}{\partial K^2}\right|_x = Z(t, T) f_{S_T|S_t}(x),$$

and thus for small $\frac{1}{\lambda}$,

$$\left.B_{K,\lambda}(t, T)\right|_x \approx \left.\frac{1}{\lambda}\frac{\partial^2 C_K(t, T)}{\partial K^2}\right|_x = \frac{1}{\lambda} Z(t, T) f_{S_T|S_t}(x).$$

The butterfly price equals the probability density times the (half-)width of the small triangle either side of $K = x$, or equivalently, the probability of S_T being in the interval $K \pm 1/\lambda$, multiplied by $Z(t, T)$. Therefore, by trading butterflies we are trading probability, in particular the risk-neutral probabilities of possible states of the random variable S_T. If a trader's assessment of the actual probability of a possible state of the world differs significantly from the risk-neutral probability of that state implied by option prices, a trade opportunity arises.

11.3 Calls as spanning set

We have shown we can move from risk-neutral probabilities to call prices via the fundamental theorem by

$$C_K(t, T) = Z(t, T) E_*((S_T - K)^+ \mid S_t).$$

And we can move from call prices to the risk-neutral density via

$$f_{S_T|S_t}(x) = \left.\frac{1}{Z(t, T)}\frac{\partial^2 C_K(t, T)}{\partial K^2}\right|_x.$$

Thus we have established an elegant duality between the set of call prices with maturity T and the risk-neutral distribution of S_T. Knowledge of call prices with maturity T determines the risk-neutral distribution at T.

Note Under the lognormal risk-neutral distribution, call prices are given by the Black–Scholes formula. We leave it as an exercise to verify that differentiating the Black–Scholes formula twice does indeed reproduce the lognormal density function.

There is an appealing dichotomy established by the two main derivative contracts we have encountered in the book so far, forwards and options. Forwards are as far one can go in developing pricing machinery without any distributional assumptions. Call options, on the other hand, tell us everything about the risk-neutral distribution at T.

Call prices do not, however, tell us about the path of S_{T^*} for $T^* < T$. For example, if we observe call prices for two maturities $T_1 < T_2$, we can determine the risk-neutral distributions for S_{T_1} and S_{T_2}, but would not be able to determine the distribution of $S_{T_2} - S_{T_1}$. Even knowing call prices $C_K(t, T^*)$ for all K and for all $T^* \leq T$ does not determine the random process for the stock price. Elegant work in stochastic processes by Dupire (1994) and Derman and Kani (1994) gives conditions for when such information does indeed determine the stochastic process for S_t.

The notion of calls as a spanning set can be seen in a variety of ways. We can picture how we might replicate any arbitrary payout $g(S_T)$ at time T with a portfolio of calls. We demonstrate this rigorously in the exercises with an elegant use of an exact Taylor series that shows that any option payout at time T can be replicated by a linear combination of the ZCB, the stock and calls with strikes $K > 0$.

We know we must be able to perform this replication since we are able to move from call prices to the price of an arbitrary derivative contract with payout $g(S_T)$ by the route

$$\text{call prices} \rightarrow \text{risk-neutral density} \rightarrow E_*\left(g\left(S_T\right)\right).$$

11.4 Implied volatility

The Black–Scholes formula is a one-to-one function between volatility and call price. In particular, given the observed call option price $C_K(t, T)$, one can determine the volatility σ_K that would give this price when entered into the Black–Scholes formula. This is known as the ***implied volatility***. Implied volatilities are calculated numerically as there is no analytic solution to the inverse of the formula.

In practice, implied volatilities for observed option prices of the same maturity but different strikes are usually different. This does not mean we have the absurd inconsistency of simultaneously requiring a lognormal distribution for S_T with one volatility σ_{K_1} for pricing call options with strike K_1, and a lognormal distribution for S_T with a different volatility σ_{K_2} to price options with strike K_2. There can, of course, only be one risk-neutral distribution for S_T. Rather, different implied volatilities are a manifestation that the risk-neutral distribution for S_T is not lognormal. (If it were lognormal, one volatility would determine option prices for all strikes.) Practitioners use implied volatilities as an efficient way to capture option prices and hence the risk-neutral distribution via

$$\{\sigma_K; K > 0\} \xleftrightarrow{\text{Black–Scholes}} \{C_K(t, T); K > 0\} \longleftrightarrow \text{risk-neutral distribution.}$$

The term ***volatility skew*** is used to describe the phenomenon that σ_K is not constant for all K, and hence that the risk-neutral distribution is not lognormal.

11.5 EXERCISES

1. **Calls as spanning set**
   (a) For any twice-differentiable positive function $g(x)$, prove by integrating the left hand side by parts that

$$\int_0^\infty (x-K)^+ g''(K)\,dK = g(x) - g(0) - xg'(0).$$

   (b) Hence, by taking risk-neutral expectations of both sides, show that the price at time t of a derivative with payout $g(S_T)$ at time T is given by

$$g(0)Z(t,T) + g'(0)\,S_t + \int_0^\infty C_K(t,T)g''(K)\,dK,$$

   where, as usual, $C_K(t,T)$ is the price of a K-strike call. The first two terms give the price of what simple portfolio that approximates the derivative payout?

2. **Recovering the risk-neutral density**
   Differentiate twice with respect to K the expression for the price of a call

$$C_K(t,T) = Z(t,T)\int_K^\infty (x-K)f(x)\,dx$$

   to find directly the risk-neutral probability density function $f(x)$ in terms of $C_K(t,T)$.

3. **Power call spanned by calls**
   Use Question 1(b) to express the price of the power call (Chapter 10 Question 3) as an integral of call prices.

PART IV
Interest Rate Options

12

Caps, floors and swaptions

Interest rate options are by far the largest options market in the world, and play a major role at the heart of the financial system. In the next few chapters, we introduce a range of widely traded interest rate derivatives, starting with European options on libor rates and swaps, known as caps, floors and swaptions. Together with swaps and forward rate agreements, these options constitute the $500 trillion notional of interest rate derivatives outstanding as of December 2011.

12.1 Caplets

A ***caplet*** struck at K on the libor rate $L_T[T, T+\alpha]$ has payout

$$\alpha \max\{L_T[T, T+\alpha] - K, 0\} \text{ at time } T+\alpha.$$

That is, a caplet is a call on a libor rate.

Here, and in Chapter 14, we will use the abbreviated notation $L_T = L_T[T, T+\alpha]$ for the libor rate, and $L_{tT} = L_t[T, T+\alpha]$ for the forward libor rate.

Let $C_K(t, T)$ be the price at time t of the caplet expiring at T with strike K. We will use the notation $C_K^{AP}(t, T)$ only when we need to distinguish from a call $C_K^{ALL}(t, T)$. We have

$$C_K(T, T) = \alpha (L_T - K)^+ Z(T, T+\alpha).$$

Note that both L_T and $Z(T, T+\alpha)$ are random variables, unknown before T.

Why is it called a caplet? Suppose a company borrows at libor from T to $T+\alpha$, and owns a K- strike caplet on L_T. Then its borrowing cost is

$$\begin{cases} \alpha L_T & \text{if } L_T \leq K \\ \alpha K & \text{if } L_T \geq K \end{cases}$$

and is thus capped at K.

Heuristically, we would like to take the discounted expected value of this payout, like we did when the random variable was S_T, the price of the stock. However, since

$$Z(T, T+\alpha) = \frac{1}{1+\alpha L_T}$$

is itself a function of L_T, the expectation is a complicated function of the random variable L_T.

Black (1976) approached the pricing problem by assuming $Z(T, T+\alpha)$ is a constant and can be taken out of the expectation (whilst keeping the libor rate L_T random), and approximating the caplet price by

$$C_K(t,T) = Z(t, T+\alpha)\, \alpha\, E(L_T - K)^+ .$$

If one assumes L_T is lognormally distributed, then one obtains the Black-76 formula. The cap market—which is deep and liquid—adopted this formula for years, but a rigorous proof that this formula was not internally inconsistent did not come until two decades later, in the work of Miltersen et al. (1997) and Brace et al. (1997).

12.2 Caplet valuation and forward numeraire

We price the caplet in an elegant manner by a careful choice of numeraire. Recall that the fundamental theorem states that there are no arbitrage portfolios $\Longleftrightarrow$ there exists a risk-neutral probability such that ratios of asset prices to the numeraire are martingales. So far we have used the money market account M_t and ZCB with price $Z(t,T)$ as numeraires, which were convenient since $M_0 = Z(T,T) = 1$.

For caplets we simplify matters by considering the ZCB with maturity $T+\alpha$ as the numeraire, so $N_t = Z(t, T+\alpha)$, motivated by the fact that, although the libor rate fixes at T, payments are made at $T+\alpha$. For the caplet, the fundamental theorem becomes

$$\frac{C_K(t,T)}{Z(t,T+\alpha)} = E_*\left(\alpha(L_T - K)^+ \mid L_{tT}\right),$$

where E_* is the risk-neutral expectation with respect to $Z(t, T+\alpha)$, known as the ***forward numeraire***. Thus we can indeed write

$$C_K(t,T) = Z(t,T+\alpha)\alpha E_*\left((L_T - K)^+ \mid L_{tT}\right)$$

under this risk-neutral distribution.

This formulation provides something else of importance. Note that

$$L_{tT} = \frac{Z(t,T) - Z(t,T+\alpha)}{\alpha Z(t,T+\alpha)},$$

that is, the forward libor is itself the ratio of assets to the numeraire, therefore by the fundamental theorem must itself be a martingale. Hence we must have

$$E_*(L_T|L_{tT}) = L_{tT},$$

the forward libor known at time t. So if we choose a lognormal risk-neutral distribution with respect to the forward numeraire, it must be of the form

$$L_T \mid L_{tT} \sim \text{lognormal}\left(\log L_{tT} - \frac{1}{2}\sigma^2(T-t),\ \sigma^2(T-t)\right),$$

under which we indeed have

$$E_*(L_T \mid L_{tT}) = L_{tT}.$$

We hence obtain rigorously the ***Black-76 formula***

$$C_K(t,T) = \alpha Z(t,T+\alpha)E_*\left((L_T-K)^+ \mid L_{tT}\right) = \alpha Z(t,T+\alpha)\left(L_{tT}\Phi(d_1) - K\Phi(d_2)\right),$$

where

$$d_1 = \frac{\log\left(\frac{L_{tT}}{K}\right) + \frac{1}{2}\sigma^2(T-t)}{\sigma\sqrt{T-t}} \text{ and } d_2 = d_1 - \sigma\sqrt{T-t}.$$

A neat choice of numeraire has allowed us to treat the libor rate very similarly to a stock. Note that if the risk-neutral distribution of L_T with respect to forward numeraire $Z(t,T+\alpha)$ is lognormal, then its risk-neutral distribution with respect to the $Z(t,T)$ numeraire may be complex. We briefly sketch how one might change from one probability distribution to another in continuous time in Chapter 16.

A ***cap*** is a portfolio of consecutive caplets. In particular, a cap from T_0 to T_n, where $T_{i+1} = T_i + \alpha$, is a series of caplets, with expiries $T_0, \dots, T_{n-1}$ and payout dates $T_1, \dots, T_n$, and is called a T_0 ***by*** T_n ***cap***. The structure of a cap is shown in Figure 12.1. Again, typically we have α to be three or six months.

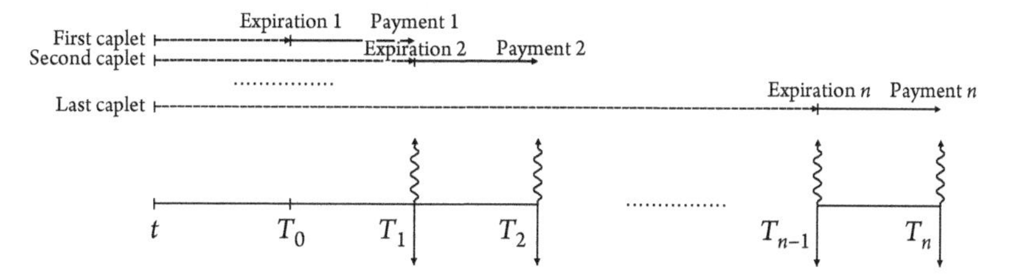

Figure 12.1 A cap

A ***floorlet*** struck at K on libor rate L_T has payout

$$\alpha \max\{K - L_T, 0\} \text{ at time } T + \alpha.$$

That is, a floorlet is a put on a libor rate. A ***floor*** is a portfolio of consecutive floorlets. A cap plus a floor of the same strike and dates is called a ***cap-floor straddle***.

12.3 Swaptions and swap numeraire

Recall from Chapter 4 that the forward swap rate

$$y_t[T_0, T_n] = \frac{Z(t, T_0) - Z(t, T_n)}{P_t[T_0, T_n]} = \frac{\sum_{i=1}^{n} L_t[T_{i-1}, T_i]\alpha Z(t, T_i)}{\sum_{i=1}^{n} \alpha Z(t, T_i)}.$$

It is complicated to determine the dynamics of $y_t[T_0, T_n]$ in terms of the dynamics of zero coupon bonds or libor rates. But we can apply a similar elegant technique as used in caplet valuation. An appropriate choice of numeraire allows us to work directly with the distribution of the swap rate $y_{T_0}[T_0, T_n]$, conditional on $y_t[T_0, T_n]$. This allows us to take the pv01 term out of the expectation and to deal with the random variable $y_{T_0}[T_0, T_n]$ analogously to L_T.

We set $T = T_0$ for ease of notation. A ***payer swaption***, struck at K with expiry T on a swap from T to T_n, is the option to pay fixed K and receive libor on the swap. This is called a ***T into $T_n - T$ payer swaption***. We display a swaption graphically in Figure 12.2.

The swaption will only be exercised at T if $y_T[T, T_n] \geq K$. If $y_T[T, T_n] < K$, then it is clearly preferable to let the swaption expire worthless rather than enter into a swap with negative value. One can always by definition pay fixed on a swap at rate $y_T[T, T_n] < K$.

A ***receiver swaption*** struck at K is the option to receive fixed K and pay libor on the swap. A ***swaption straddle*** is a payer and a receiver swaption of the same strike and dates.

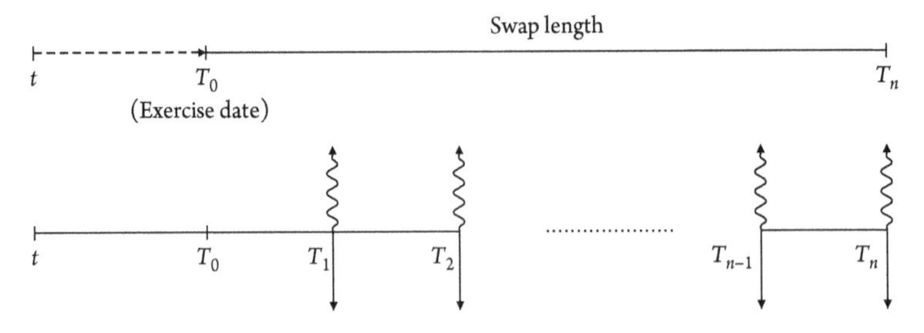

Figure 12.2 A swaption

If we denote the price of the payer swaption at time $t \leq T$ by $\Psi_K(t, T, T_n)$, then its payout at exercise date T is given by

$$\Psi_K(T, T, T_n) = (y_T[T, T_n] - K)^+ P_T[T, T_n].$$

A payer swaption can be thought of heuristically as a call on a swap rate, multiplied by the pv01 term $P_T[T, T_n]$.

Note Here, the exercise date and start date of the swap coincide. This holds for the vast majority of swaptions traded in the market, although does not have to be the case. For example, one could consider a swaption with exercise date T on a swap from time T^* to T_n, where $T^* > T$. These are called ***midcurve swaptions***.

The swaption payout is similar to the caplet payout, with the swap rate analogous to the libor rate, and the swap pv01 analogous to the ZCB with maturity $T + \alpha$. We will apply a similar approach to valuation. We pick $P_t[T, T_n]$ as the numeraire, which is a positive asset being a sum of ZCBs. By the fundamental theorem,

$$\frac{\Psi_K(t, T, T_n)}{P_t[T, T_n]} = E_* \left(\frac{\Psi_K(T, T, T_n)}{P_T[T, T_n]} \right) = E_* (y_T[T, T_n] - K)^+,$$

where E_* is the expectation under the risk-neutral distribution with respect to $P_t[T, T_n]$. This is called the ***swap numeraire***.

Once again, $y_t[T, T_n]$ is itself a ratio of assets to the numeraire, so must itself be a martingale under the risk-neutral distribution. If we choose a lognormal risk-neutral distribution, then we must have

$$y_T[T, T_n] \mid y_t[T, T_n] \sim \text{lognormal}\left(\log y_t[T, T_n] - \frac{1}{2}\sigma^2(T-t), \sigma^2(T-t) \right),$$

and hence we obtain the ***Black formula for swaptions***

$$\Psi_K(t, T, T_n) = P_t[T, T_n] \left(y_t[T, T_n]\Phi(d_1) - K\Phi(d_2) \right)$$

where d_1 and d_2 are exactly as in the Black-76 formula, with $y_t[T, T_n]$ replacing $L_t[T, T + \alpha]$.

As mentioned in Chapter 11, there is no reason why the risk-neutral distribution for the swap rate or libor need be lognormal. Similar to the Black–Scholes formula, the Black-76 formula is a one-to-one function between volatility and option price. The presence of a volatility skew would indicate that the risk-neutral distribution is not lognormal.

Note The simple notation T for the expiration does not capture the vital importance of the exact expiration date and time for a swaption trader. Throughout the 1990s, most swaptions were exercised by voice, meaning the owner of the swaption would need to call the seller of the swaption to inform them of exercise. The expiration of almost all US dollar swaptions is at 11am New York time, and every options trader would have a visceral fear of forgetting to exercise, of looking at the clock and seeing 11.05am. Rare errors would typically be resolved between bank management, as both sides knew the situation could be reversed at some point in the future. Now, virtually all swaptions and other options are auto-exercised, meaning that they are considered to have been exercised (whether physical or cash settlement) if at the exercise date the option is in the money—for example, if $y_T[T, T_n] > K$ in the case of a payer swaption.

12.4 Summary

We summarize in Table 12.1 the interest rate options encountered so far, using the abbreviations $L_T = L_T[T, T+\alpha]$ and $y_T = y_T[T, T_n]$.

Table 12.1 Summary of interest rate options.

Name	Description	Payout at T	Usual numeraire
FRA		$\alpha(L_T - K)Z(T, T+\alpha)$	
Caplet	Call on L_T	$\alpha(L_T - K)^+Z(T, T+\alpha)$	$Z(t, T+\alpha)$
Floorlet	Put on L_T	$\alpha(K - L_T)^+Z(T, T+\alpha)$	$Z(t, T+\alpha)$
Cap	Series of caplets		Multiple $Z(t, T_i+\alpha)$
Cap-floor straddle	Cap + floor		Multiple $Z(t, T_i+\alpha)$
Swap (pay fixed)		$(y_T - K)P_T[T, T_n]$	
Payer swaption	Call on y_T	$(y_T - K)^+P_T[T, T_n]$	$P_t[T, T_n]$
Receiver swaption	Put on y_T	$(K - y_T)^+P_T[T, T_n]$	$P_t[T, T_n]$
Swaption straddle	Payer + receiver	$\mid K - y_T \mid P_T[T, T_n]$	$P_t[T, T_n]$

12.5 EXERCISES

1. **Black-76 formula**
   Derive the Black-76 formula. That is, suppose the risk-neutral distribution under the forward numeraire is given by

$$L_T \mid L_{tT} \sim \text{lognormal}\left(\log L_{tT} - \frac{1}{2}\sigma^2(T-t),\ \sigma^2(T-t)\right).$$

Use the fundamental theorem to prove that the price of a K-strike caplet is given by

$$C_K(t,T) = \alpha Z(t,T+\alpha)\left(L_{tT}\Phi(d_1) - K\Phi(d_2)\right),$$

where

$$d_1 = \frac{\log\left(\frac{L_{tT}}{K}\right) + \frac{1}{2}\sigma^2(T-t)}{\sigma\sqrt{(T-t)}} \text{ and } d_2 = d_1 - \sigma\sqrt{(T-t)}.$$

2. **Digital caplets**
   Consider two possible candidates (I) and (II) for the risk-neutral distribution of L_T conditional on L_{tT}. Both are distributions with respect to the numeraire $Z(t,T+\alpha)$.
   (I) $L_T \mid L_{tT} \sim \text{lognormal}\left(\log L_{tT} - \frac{1}{2}\sigma^2(T-t), \sigma^2(T-t)\right)$.
   (II) $L_T \mid L_{tT} \sim N(\mu, \psi^2(T-t))$.
   (a) Assuming no-arbitrage, what is the value of μ?
   (b) Calculate the price at time t of the digital caplet that pays α at time $(T+\alpha)$ if $L_T > K$, and zero otherwise, under the two different models (I) and (II).
   (c) Calculate the vega of this digital caplet under the two models (I) and (II) — where vega is defined as $\frac{\partial}{\partial\psi}$ for (II). For what values of K does the digital caplet have zero vega, for (I) and (II) respectively?
   (d) What are the prices of the at-the-money-forward digital caplet for (I) and (II)? Explain your answers in terms of the median of lognormal and normal distributions.

13

Cancellable swaps and Bermudan swaptions

One of the most important applications of interest rate options within financial markets are cancellable swaps. Cancellable swaps can be European style with just one cancellation date, Bermudan style with multiple cancellation dates, or American style with continuous cancellation dates (which in practice usually means daily). We here outline the construction of cancellable swaps using swaptions, and explore several features of Bermudan swaptions.

13.1 European cancellable swaps

Suppose we pay fixed K versus receiving libor on a swap from $T = T_0$ to T_n, and we have the option to cancel at a single time $T_j > T$ for no cost. That is, if we cancel the swap at T_j, no subsequent swap payments are made and thus the value at T_j is trivially zero. Therefore, we would cancel the swap at time T_j if and only if the swap rate $y_{T_j}[T_j, T_n] < K$, since otherwise we would retain a swap of negative value.

We can deconstruct the cancellable swap into a vanilla swap with fixed rate K from T to T_n, plus a T_j into $T_n - T_j$ receiver swaption with strike K. The option to cancel the swap is precisely achieved by exercising the receiver swaption, which exactly offsets the swap from T_j to T_n.

The cancellable swap can also be deconstructed into a swap with fixed rate K from T to T_j, plus a T_j into $T_n - T_j$ payer swaption with strike K. Using this formulation, the act of not cancelling the swap is the same as deciding to exercise the payer swaption, both of which are done if and only if $y_{T_j}[T_j, T_n] \geq K$.

The equivalence of these two constructs, which are shown in Figure 13.1, also follows from put-call parity for swaps. At time $t \leq T_j$, a swap where one pays a fixed rate K from T_j to T_n is equal in value to being long a T_j into $T_n - T_j$ K-strike payer swaption, and short a T_j into $T_n - T_j$ K-strike receiver swaption.

If the fixed rate K on the cancellable swap is equal to $y_T[T, T_n]$, the par swap rate at T for period T to T_n, then the cancellable swap must have positive value at T, since the

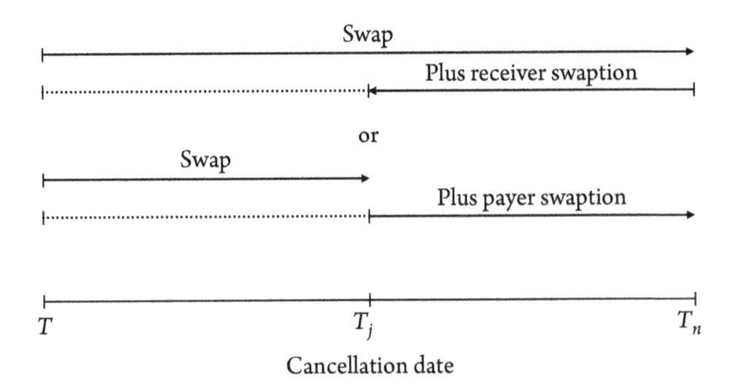

Figure 13.1 Deconstruction of a cancellable swap

swap from T to T_n has zero value by definition and the T_j into $T_n - T_j$ receiver swaption has positive value (except in the negligible case that volatility is zero and the receiver is out-of-the-money).

The following question naturally arises. Is there a value K^* of K for which the cancellable swap has zero value at T? Using our second deconstruction, the value of the cancellable swap is the sum of the value of the swap from T to T_j and the price of the T_j into $T_n - T_j$ payer swaption. If $K = y_T[T, T_j]$, then the swap has zero value and the swaption has positive value, so the cancellable swap has positive value. If $K > y_T[T, T_j]$, then the swap has negative value. As we let $K \to \infty$, the price of the payer swaption decreases to zero.

A mathematician can now appeal to the intermediate value theorem, which states there must exist $K^* > y_T[T, T_j]$ such that the cancellable swap has zero value. Note from above that we also must have $K^* > y_T[T, T_n]$. In the case of spot-starting swaps where $T = 0$, we call K^* the ***T_n noncall T_j (European cancellable) swap rate***, 'noncall' indicating it is not cancellable until T_j.

13.2 Callable bonds

Corporations issuing fixed rate debt often use swaps to exchange the fixed rate payments they are making on the bond for floating libor payments. It might be the case, for example, that investors prefer fixed rate bonds to floating rate bonds, but the corporate issuer prefers floating rate liabilities. By swapping a fixed rate bond as shown in Figure 13.2, the corporation ends up paying an interest cost of libor plus or minus a spread. The fixed rate of the

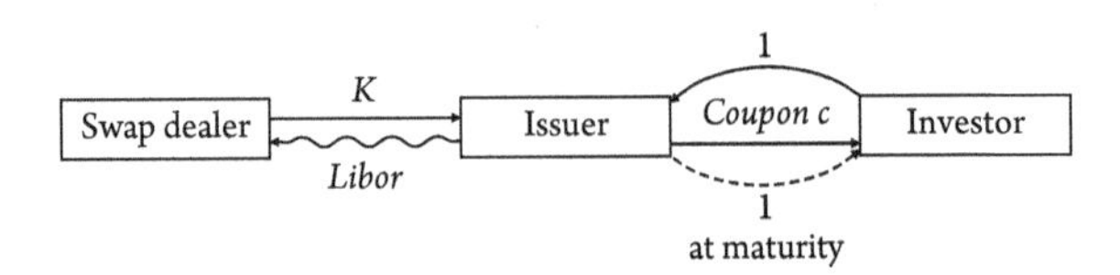

Figure 13.2 Swapping a fixed rate bond

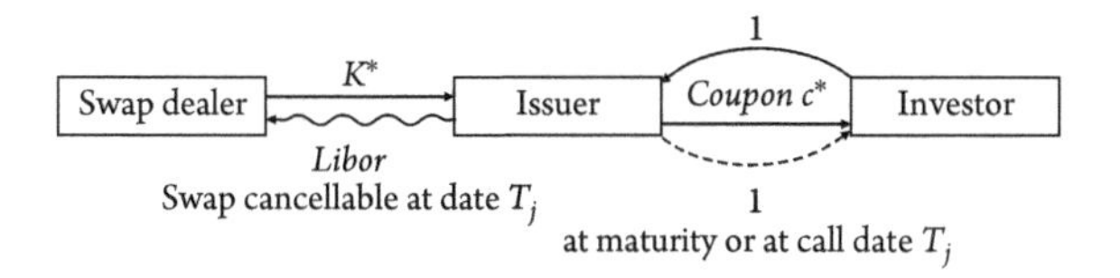

Figure 13.3 Swapping a callable bond

bond is determined both by market interest rates and investor preference for the credit of the issuer. The swap rate, however, is essentially independent of the credit of the issuer.

The company may alternatively issue a callable bond, which they have the option to buy back (or 'call') from the investor at 100% of the notional (or 'par') at a fixed time T_j before maturity. In a similar manner to above, the company often will exchange these fixed payments for libor payments using a cancellable swap, as shown in Figure 13.3. The swap counterparty now has the right to cancel the swap at T_j. Typically, the issuer will automatically call the bond if the swap is cancelled.

Why would a company issue a callable bond with coupon c^* as opposed to a regular non-callable (also known as 'bullet') bond with coupon c? Indeed, why are callable bonds such a popular instrument both for issuers and investors?

Looking first at the economics of the two trades, suppose, as is typically the case, that both swaps are executed at time T at current market levels, so $K = y_T[T, T_N]$ and K^* is the cancellable swap rate. Then the net funding cost for the company for the bullet bond is libor $+ (c - K)$ and for the callable bond is libor $+ (c^* - K^*)$. The additional coupon $(c^* - c)$ an investor receives for buying the callable bond is often less than the difference $(K^* - K) = (K^* - y_T[T, T_N])$ between the cancellable and regular swap rates. The former incorporates investor preference for the two different types of bond, whilst the latter is based on the market price for swaptions. Whilst investors may often want a bond with a higher coupon and so are attracted to the callable version, they may not have properly assessed the extra coupon they should receive for selling the cancellation option.

Note The optics of a callable bond can appear attractive to investors. For example, suppose a company can issue a two-year bond with a coupon of 5.00%, a five-year bond with a coupon of 5.20%, and a five-year noncall two-year European callable bond with a coupon of 5.45%, each bond being issued at par. The investor might conclude that the callable bond is attractive, since the investor beats the two-year rate 5.00% if the bond is called after two years, and beats the five-year rate 5.20% if the bond is not called and left outstanding for five years. However, if rates go down sufficiently for the bond to be called after two years, the investor receives 5.45% for two years, but then needs to reinvest the notional at the prevailing low rate. The investor thus bears significant reinvestment risk with the callable bond.

In practice, it might be that the two-year swap rate is 4.50%, the five-year swap rate is 4.70% and the five-year noncall two-year rate is 5.00%. Then the issuer has improved their funding cost from libor+50bp to libor+45bp by issuing the callable bond.

13.3 Bermudan swaptions

We now introduce the Bermudan swaption, which is my favourite interest rate option. Bermudans have an appealing combination of being easy to define, and yet possessing unexpected dimensions of subtlety. Pricing Bermudans in all their complexity is a rich field beyond our scope and still the subject of current research. In this book we develop a broad range of arbitrage bounds and model-independent results for Bermudans, which we can derive without advanced option pricing machinery.

Let us first recap interest rate derivatives we have encountered so far.

A swap is an exchange of cashflows where one counterparty pays a fixed rate K and receives libor each period from T_0 to T_n.

A T_0 into $T_n - T_0$ European payer swaption with strike K is the option at T_0 to pay K and receive libor from T_0 to T_n.

A T_0 by T_n cap with strike K can be defined as a portfolio of options at each T_i to pay αK and receive libor $\alpha L_{T_i}[T_i, T_i + \alpha]$ at each $T_i + \alpha$. If $T_0 = 0$, this is simply called a T_n-cap.

The libor rates referenced by a T_0 by T_n cap are the same as those referenced by a T_0 into $T_n - T_0$ swaption. For example, a two-year into three-year (2yr-into-3yr) swaption references the same libor rates as a two-year by five-year (2yr-by-5yr) cap.

Note A cap is sum of options on the libor rates $L_{T_i}[T_i, T_i + \alpha]$, whilst a swaption is an option on a weighted average of the same $L_{T_i}[T_i, T_i + \alpha]$. The relative value between these instruments forms a significant component of the interest rate derivative trading landscape.

A ***T_0 into $T_n - T_0$ Bermudan payer swaption*** is the option at each T_i to pay K and receive libor from T_i to T_n, for $i = 0, \ldots, n-1$. If the swaption is not exercised at T_i, then the option continues. Once the option is exercised at a particular T_i, all subsequent options disappear. That is, there are many exercise dates but only one option.

Figure 13.4 shows the structure of a European swaption, a cap and a Bermudan swaption. We now establish a series of bounds on the price of the Bermudan in terms of (European-style) interest rate options.

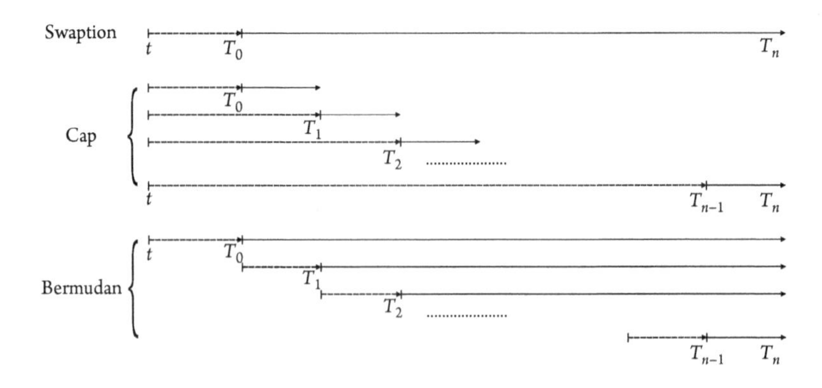

Figure 13.4 A swaption, a cap and a Bermudan swaption

Let $\Psi_K(t, T_j, T_n)$ be the price at time t of the K-strike European payer swaption, with exercise date T_j on swap from T_j to T_n.

Let $C_K(t, T_0, T_n)$ be the price at time t of the T_0 by T_n K-strike cap.

Let $B_K(t, T_0, T_n)$ be the price at time t of the K-strike Bermudan payer swaption, with first exercise date T_0 on a swap ending at T_n.

We will assume for simplicity that the payment frequencies on all derivatives are quarterly, and hence our interest accrual factor $\alpha = 0.25$. In other words, we assume the cap is on three-month libor, and that all swaps have quarterly payments on the fixed side versus three-month libor quarterly payments on the floating side. (In practice, the frequency of the various options may vary.)

Result

$$(a)\ C_K(t, T_0, T_n) \geq B_K(t, T_0, T_n) \geq \max_{0 \leq i \leq n-1} \Psi_K(t, T_i, T_n).$$

$$(b) \sum_{0 \leq i \leq n-1} \Psi_K(t, T_i, T_n) \geq B_K(t, T_0, T_n).$$

Result (a) shows that the Bermudan payer swaption can in some sense be considered as the intermediate option between the European payer swaption and the cap.

Proof To prove an inequality, we assume it does not hold, and show that we can then construct an arbitrage portfolio.

For example, we prove the second inequality in (a) as follows. Suppose $B_K(t, T_0, T_n) < \Psi_K(t, T_j, T_n)$ for some $0 \leq j \leq n-1$. Then we can sell the T_j into $T_n - T_j$ European payer swaption and buy the Bermudan swaption, receiving a positive amount of cash. Equivalently the portfolio consisting of short one European swaption, long one Bermudan swaption and a positive holding of cash has zero value at time t.

If the European is not exercised at time T_j, we do not exercise the Bermudan. We are left with a positive amount of cash. If the European is exercised at time T_j, we exercise the Bermudan at time T_j, setting up two exactly offsetting swaps. Once again, the portfolio is left with a positive amount of cash. Hence we have created an arbitrage portfolio. ■

The other results in (a) and (b) follow in a similar manner. This form of argument is a powerful application of the no-arbitrage assumption, and widely used to prove results of the form $A \geq B$. To do so, we assume $A < B$; we sell at B and buy for A; and we show that this portfolio can never have negative value.

Note In 2009 during the aftermath of the Lehman Brothers bankruptcy, Bermudan receiver swaptions traded at a higher price than libor floors of the same strike and underlying dates. This meant that the elementary arbitrage bounds on Bermudan swaption prices developed here and in Question 2(a) were violated. The episode was a particularly vivid example of derivative prices during the financial crisis challenging solid no-arbitrage arguments that practitioners had long taken for granted. Whilst the arbitrage bound violation lasted only a few days, it did nevertheless provide a cautionary note regarding the foundations of derivative pricing. See Blyth (2012) for further discussion.

13.4 Bermudan swaption exercise criteria

Consider the decision at time T_i whether or not to exercise the Bermudan swaption.

If we exercise the Bermudan payer at T_i, we have a swap with value $(y_{T_i}[T_i, T_n] - K)P_{T_i}[T_i, T_n]$.

If we do not exercise, we have a Bermudan swaption with first exercise T_{i+1} and with value $B_K(T_i, T_{i+1}, T_n)$.

Therefore, the exact exercise condition at time T_i, as shown in Figure 13.5, is given by

$$\text{exercise Bermudan at } T_i \iff (y_{T_i}[T_i, T_n] - K)P_{T_i}[T_i, T_n] > B_K(T_i, T_{i+1}, T_n).$$

This is of limited use since we do not yet have a way of valuing the subsequent Bermudan option. In practice, we would attempt to build a tree for the evolution of interest rates and work backwards through the tree, in a similar manner to how we priced the American put on a stock on the binomial tree (see Question 1 in Chapter 8).

However, we can make progress by considering model-independent exercise criteria, which depend only upon swap rates and European swaption prices observed at the exercise date. We have three tiers of model-independent criteria, all of which give conditions on when not to exercise the Bermudan.

Result *Do not exercise if* $y_{T_i}[T_i, T_n] < K$.

This is obvious since the subsequent Bermudan has non-negative value $B_K(T_i, T_{i+1}, T_n)$.

Result *We know that*

$$B_K(T_i, T_{i+1}, T_n) \geq \max_{i+1 \leq j \leq n-1} \Psi_K(T_i, T_j, T_n).$$

Therefore, do not exercise if

$$(y_{T_i}[T_i, T_n] - K)\, P_{T_i}[T_i, T_n] \leq \max_{i+1 \leq j \leq n-1} \Psi_K(T_i, T_j, T_n).$$

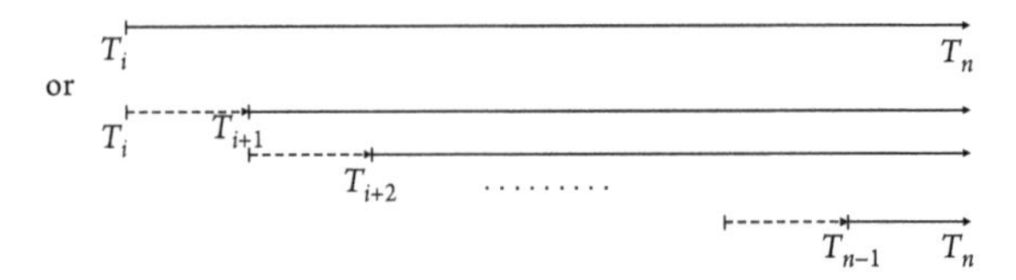

Figure 13.5 Bermudan exercise decision

That is, we do not exercise if the intrinsic value (the value of the underlying swap) is less than the most valuable of the subsequent European swaptions.

In particular, the Bermudan exercise criterion, and hence its value, depends on the 'co-terminal' or 'diagonal' swaptions with prices $\Psi_K(T_i, T_j, T_n)$, for $j = i + 1, \ldots, n - 1$. For example, the price of a one-year into nine-year Bermudan swaption depends on the price of the one-year into nine-year, two-year into eight-year, . . ., and nine-year into one-year European swaptions.

There is a further interesting model-independent exercise criterion, which does not depend in any way on option prices or levels of volatility.

Result *Do not exercise the Bermudan payer at T_i if*

$$y_{T_i}[T_i, T_j] < K \textit{ for any } i + 1 \leq j \leq n.$$

Proof Suppose $y_{T_i}[T_i, T_k] < K$ for some k and consider two alternative strategies.

A: Exercise the Bermudan payer at T_i.

B: Pay the fixed rate $y_{T_i}[T_i, T_k]$ and receive libor from T_i to T_k, then exercise the Bermudan at T_k.

As shown in Figure 13.6, strategy **B** dominates strategy **A**, in the sense it always produces a portfolio of greater value. Therefore, it can not be optimal to exercise at T_i. ■

The last result shows that the exercise decision for the Bermudan depends not only on the diagonal swaptions, but also on par swap rates to all intermediate exercise dates. Often models for pricing the Bermudan option do not explicitly incorporate the latter. We have several examples of Bermudan exercise criteria in the exercises.

Note I once interviewed a finance professional who had run a Bermudan trading book at a major firm for many years. He refused to believe that the exercise criteria for a Bermudan involved the level of swap rates to intermediate dates, insisting that only the diagonal swaptions were relevant. This episode showed that practitioners can sometimes get so involved in their markets and their sophisticated pricing models that relatively straightforward, model-independent results are overlooked.

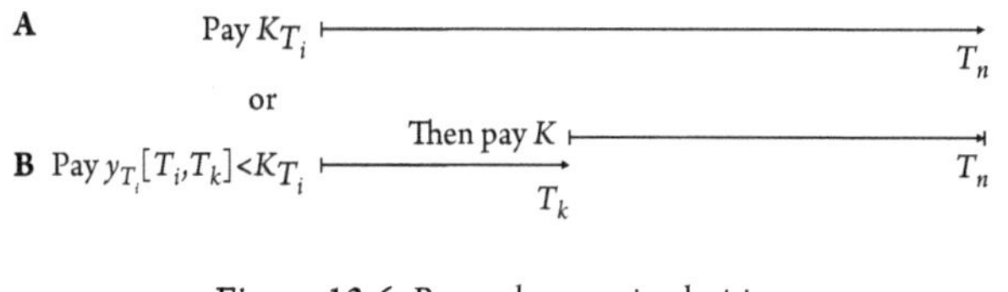

A Pay K_{T_i} ⟶ T_n

or

B Pay $y_{T_i}[T_i,T_k]<K_{T_i}$ ⟶ T_k, Then pay K ⟶ T_n

Figure 13.6 Bermudan exercise decision

13.5 Bermudan cancellable swaps and callable bonds

The concept of the cancellable swap extends naturally from the European to the Bermudan case. Suppose one pays fixed K and receives libor from T_0 to T_n, with rights to cancel at T_i, for $i = j, \ldots, n-1$. For spot-starting swaps where $T_0 = 0$, we call this a T_n ***noncall*** T_j ***Bermudan cancellable swap.***

The Bermudan cancellable swap can be deconstructed into paying fixed K and receiving libor on a swap from T_0 to T_n, plus owning a T_j into $T_n - T_j$ Bermudan receiver swaption with strike K. Cancelling the swap at some time T_i is equivalent to exercising the Bermudan receiver at time T_i, which would exactly offset future swap payments. If one does not cancel at T_i, or equivalently does not exercise the Bermudan receiver, the options continue.

Note that in contrast to the European case, we cannot deconstruct the Bermudan cancellable swap using a swap where we pay fixed from T_0 to T_j, plus a T_j into $T_n - T_j$ Bermudan payer swaption. To see this, suppose we are paying fixed on the cancellable swap and at time T_j we decide to continue paying fixed—we would certainly do so, for example, if $y_{T_j}[T_j, T_n] > K$. In the construct using a Bermudan payer swaption, we would need to exercise the Bermudan at time T_j, leaving no subsequent options to cancel. More generally, there is no put-call parity for Bermudans, similar to there being no put-call parity for an American option on a stock.

As with the European cancellable swap, for the Bermudan there is a fixed rate K^{**} for which the cancellable swap has zero value. There is no established convention for the notation for this rate, given it should incorporate the current time t, the start and end date of the swap, plus the start and frequency of the cancellation options. Here we adopt the notation '5nc2 Berm s/a' to denote a five-year swap where one pays fixed with the Bermudan right to cancel after two years and semi-annually thereafter, and K{5nc2 Berm s/a} to be the cancellable swap rate, that is, the fixed rate such that the cancellable swap has zero value. Similarly, K{5nc3 Euro} is the cancellable swap rate for a five-year swap, where one pays fixed with the European right to cancel after three years.

We can establish hierarchies for ordering Bermudan and European cancellable swap rates. The following illustrative examples give the key features.

Result

(i) *K{6nc2 Berm q} ≥ K{5nc2 Berm q} ≥ K{5nc2 Berm s/a} ≥ K{5nc3 Berm s/a} ≥ K{5nc3 Euro} ≥ five-year swap rate*

(ii) *K{5nc3 Euro} ≥ three-year swap rate*

(iii) *K{5nc2 Berm s/a} ≥ K{5nc2 Euro} ≥ two-year swap rate.*

The intuition to these orderings is that one is prepared to pay a higher fixed rate for more cancellation optionality. An ordering can be established if one cancellable swap contains options to cancel which are a subset of another. Note, however, that it can be the case, for example, that

K{5nc2 Berm s/a} < K{5nc3 Berm q}, or K{5nc3 Euro } < two-year swap rate.

To prove the results, assume they do not hold and construct an arbitrage portfolio. We give one proof as an example.

Proof Let K_1 = K{5nc2 Berm s/a} and K_2 = K{5nc3 Berm s/a}. To prove $K_1 \geq K_2$, we suppose $K_1 < K_2$ and construct an arbitrage portfolio. In particular, we construct the portfolio where we pay fixed K_1 versus libor on a 5nc2 Berm s/a swap, and we receive fixed K_2 versus libor on a 5nc3 Berm s/a swap. By definition we can do this at no cost. On the latter swap it is the counterparty, the payer of the fixed rate, who has the options to cancel.

If the 5nc3 swap is never cancelled by the counterparty, we do not cancel the 5nc2 swap, and receive $K_2 - K_1 > 0$ for five years. If, however, the 5nc3 swap is cancelled at some time $T_j \geq 3$, we cancel the 5nc2 at the same time (we are able to do this as the options in the 5nc3 are a subset of those in the 5nc2). In this case we receive $K_2 - K_1 > 0$ for T_j years. Thus we have constructed an arbitrage portfolio. ■

Bermudan callable bonds are particularly attractive to investors since they typically have a higher coupon than European callable or noncallable bonds. The investor has sold a Bermudan option to the company in return for a higher fixed rate. The company typically does not want to manage the option exercise risk so enters into a cancellable swap with a bank as shown in Figure 13.7.

Similarly to the European case, the coupon on the Bermudan callable bond c^{**} is determined by how much the investor is willing to receive on a callable bond of that company. The difference between c^{**} and a rate on a fixed rate bond can often be less than the difference between the equivalent cancellable swap rate K^{**} and the par swap rate. Investors often have a preference for callable bonds since they have higher coupons and are prepared to accept less than the full value of the optionality.

A ***mortgage*** is another type of callable bond central to the financial system. Simplifying the structure significantly, in particular ignoring principal payments, amortization and any prepayment penalties, a mortgage can to an approximation be considered a purchase of a callable bond by the lender from the homeowner. The homeowner owns a Bermudan option embedded in their mortgage, namely the right to prepay the mortgage, or equivalently owns the right to 'call' the bond back from the lender at par. Given the homeowner is paying a fixed rate on the mortgage, their option to prepay (or cancel) such an agreement is equivalent to a Bermudan receiver swaption.

There is an extensive fixed rate mortgage market in the US, and institutions trading mortgages, such as the government sponsored enterprises Freddie Mac and Fannie Mae, often

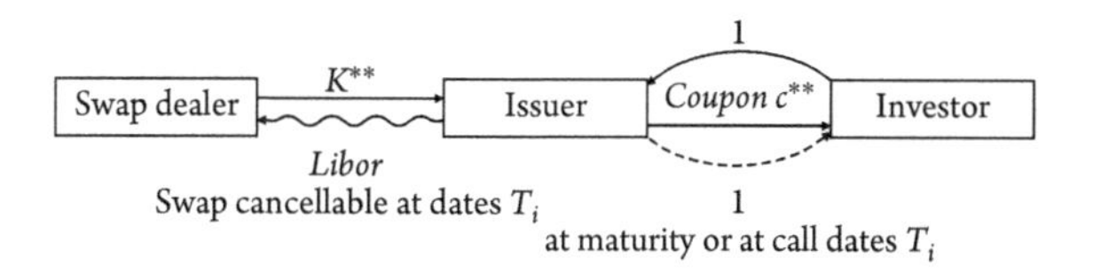

Figure 13.7 Swapping a Bermudan callable bond

use Bermudan swaptions to hedge the optionality embedded in the mortgages. As a result, Bermudan swaptions are particularly widely traded in the US and constitute an important component of derivative markets.

13.6 EXERCISES

1. **Interest rate options quick quiz**

   (a) Rank the following options in order of value, from greatest to least.

   A. 7yr 5% floor on 3mL.
   B. 7yr 5.5% floor on 3mL.
   C. 10yr 5.5% floor on 3mL.
   D. 7yr 5% floor on 3mL, the whole deal knocking out if 3mL ever fixes below 3%.
   E. 7yr 5.25% floor on 3mL, knocking out for periods where 3mL fixes above 7%.
   F. 7yr 5% floor on 3mL, knocking out for periods where 3mL fixes below 3%.

   (b) Rank the following options in order of value, from greatest to least.

   A. 2yr-into-3yr 6% payer swaption.
   B. 2yr-by-5yr 6% cap on 3mL.
   C. 2yr-into-3yr 6% Bermudan payer swaption with quarterly exercises.
   D. 2yr-into-3yr 6% Bermudan payer swaption with semi-annual exercises.
   You can assume the swaps underlying A, C and D have quarterly payment dates versus 3mL.

   (c) A customer wants to pay fixed for 15yrs, with the one-time right to cancel the swap after 5yrs. Which of the following strategies will precisely accomplish this? You may choose none, one or more than one.

   I Customer pays fixed for 15yrs and buys a 5yr-into-10yr European payer swaption.
   II Customer pays fixed for 5yrs and buys a 5yr-into-10yr European payer swaption.
   III Customer pays fixed for 15yrs and buys 5yr-into-10yr European receiver swaption.
   IV Customer pays fixed for 5yrs and buys a 5yr-into-10yr European receiver swaption.

   (d) A customer wants to have the same trade as in (c), but with Bermudan cancellation options starting in five years. Which of the following strategies will precisely accomplish this? You may choose none, one or more than one.

   I Customer pays fixed for 15yrs and buys a 5yr-into-10yr Bermudan payer swaption.

II Customer pays fixed for 5yrs and buys a 5yr-into-10yr Bermudan payer swaption.

III Customer pays fixed for 15yrs and buys a 5yr-into-10yr Bermudan receiver swaption.

IV Customer pays fixed for 5yrs and buys a 5yr-into-10yr Bermudan receiver swaption.

2. Bermudan exercise criteria

Let $B_K^r(t, T_0, T_n)$ be the price at time t of a K-strike Bermudan receiver swaption with first exercise date T_0, last exercise date T_{n-1}, and underlying swap maturity T_n.

(a) Prove by no-arbitrage arguments that

$$FLR_K(t, T_0, T_n) \geq B_K^r(t, T_0, T_n) \geq \max_{0 \leq i \leq n-1} \Psi_K^r(t, T_i, T_n),$$

where FLR_K is the price of a floor and Ψ^r is the price of a European receiver swaption.

(b) By specifying a dominating exercise strategy, prove that one should not exercise the Bermudan receiver at time T_j if

$$y_{T_j}[T_j, T_k] > K \text{ for some } k,\ j+1 \leq k \leq n.$$

Does the converse hold? That is, if $y_{T_j}[T_j, T_k] < K$ for all k, $j+1 \leq k \leq n$, should one necessarily exercise the Bermudan at T_j?

3. Ordering Bermudan swaptions

Consider the following six interest rate options, where $0 < K^* < K$. You may assume that all swaps have the same frequency, fixed quarterly versus three-month libor quarterly.

I. A 1yr-into-10yr K-strike European payer swaption, where Bank A has the right to exercise at $T = 1$ into a swap from $T = 1$ to $T = 11$.

II. A 1yr-into-10yr K-strike Bermudan payer swaption, where Bank A has the right to exercise at $T = 1$ and annually thereafter, into a swap with final maturity $T = 11$.

III. A 1yr-into-10yr K-strike Bermudan payer swaption, where Bank A has the right to exercise at $T = 1$ and semi-annually thereafter, into a swap with final maturity $T = 11$.

IV. A 1yr-into-10yr K^*-strike European payer swaption.

V. A 1yr-into-10yr K-strike American style payer swaption defined as follows. Bank A has the right to exercise at any time up to $T = 1$. If Bank A exercises at time T_0, $0 \leq T_0 \leq 1$, Bank A pays fixed on a swap from T_0 to $T_0 + 10$.

VI. A 1yr-into-10yr K-strike American style payer swaption defined as follows. Bank A has the right to exercise at any time up to $T = 1$. If Bank A exercises at time T_0, $0 \leq T_0 \leq 1$, Bank A pays fixed on a swap from $T = 1$ to $T = 11$.

For each pair of options (a)-(g) in Table 13.1 below, determine the best relationship between their values to Bank A at time $t = 0$ from choices $=, \geq, \leq$ and ?, where ? means the relationship is indeterminate.

Table 13.1 Ordering Bermudan swaptions.

		$=, \geq, \leq$ **or ?**	
(a)	I		II
(b)	II		III
(c)	I		IV
(d)	II		V
(e)	V		VI
(f)	I		VI
(g)	III		IV

4. **Bermudan exercise**

(a) Suppose you own a 1yr-into-5yr Bermudan receiver swaption with strike 6%. The underlying swap has quarterly payment dates, fixed versus three-month libor. The exercise dates are quarterly. In one year's time you must determine whether to exercise the first option to receive fixed on a five-year swap, based on the then current interest rate scenario. For each scenario A-F in Table 13.2 for interest rates in one year's time, determine the most appropriate response to the question 'Should the first option be exercised?' from the following choices.

I Yes, the option should definitely be exercised.

II No, the option should definitely not be exercised.

III It depends upon current levels of other market variables such as European swaption prices.

Table 13.2 Exercise decision.

Scenario	3mL %	5yr swap %	I, II or III?
A	6.1	6.1	
B	6.1	5.9	
C	6.1	3.5	
D	3.5	6.1	
E	3.5	5.9	
F	3.5	3.5	

(b) Suppose you are told volatilities are zero for all interest rate options, that is, all out-of-the-money-forward options have zero value. Again answer the question 'Should the first option be exercised?' for scenarios D, E and F in Table 13.3.

Table 13.3 Revised exercise decision.

Scenario	3mL %	5yr swap %	I, II or III?
D	3.5	6.1	
E	3.5	5.9	
F	3.5	3.5	

5. **Double-cancellable swaps**

Recall that K{10nc *x* Berm ann} is the fixed rate such that a swap, where we pay fixed for ten years with the right to cancel after *x* years and annually thereafter, has zero value. Suppose you observe the following rates.

K{10nc1 Berm ann} 7.20%.

K{10nc2 Berm ann} 7.10%.

One-year swap rate 6.10%.

Two-year swap rate 6.15%.

Ten-year swap rate 7.00%.

(a) Bank A wants to pay fixed on a {10nc1 Berm ann} swap to Bank B. However, suppose in addition that Bank B itself has the European option to cancel this swap after one year. Determine the most likely rate for this 'double-cancellable' swap from the following.

I $> 7.20\%$

II 7.20%

III $> 6.10\%$ but $< 7.20\%$

IV 6.10%

V $< 6.10\%$

Give your reasoning.

(b) Bank A wants to pay fixed on a {10nc2 Berm ann} swap to Bank B, and once again Bank B has the European option to cancel the swap after one year. Determine the most likely rate for this 'double-cancellable' swap from the following.

I $> 7.10\%$

II 7.10%

III $> 6.15\%$ but $< 7.10\%$

IV 6.15%

V $< 6.15\%$

14

Libor-in-arrears and constant maturity swap contracts

In this chapter we discuss libor-in-arrears and more general convexity corrections. Libor-in-arrears is one of the simplest members of the universe of ***exotic***, or non-standard, interest rate derivatives. Whilst easy to define, it involves subtle pricing methodology.

14.1 Libor-in-arrears

Early in Chapter 1 we discussed that finance has a habit of giving rise to unexpected complexity in what ostensibly appear to be simple areas. Specifically, we mentioned in Chapter 3 that changing the time at which cashflows from a FRA are paid, simply from time $T + \alpha$ to time T, opens up a new realm of subtle pricing problems. We explore this issue here, using the classic example of libor-in-arrears. Although easily defined, this is a topic that often confuses.

We again use the abbreviated notation $L_T = L_T\left[T, T + \alpha\right]$ and $L_{tT} = L_t\left[T, T + \alpha\right]$, for $t \leq T$.

Recall that a regular FRA is an agreement where the buyer of the FRA pays αK and receives αL_T, all paid at time $T + \alpha$ (usually with α being three or six months). Thus the value of the FRA at time T is

$$\alpha(L_T - K)Z\left(T, T + \alpha\right) = 1 - \left(1 + \alpha K\right)Z\left(T, T + \alpha\right).$$

We saw from Chapter 3 that the FRA can be replicated by a holding in ZCBs and a libor deposit, and hence the forward libor rate L_{tT}, the value of K such that the FRA has zero value at time t, is given by

$$L_{tT} = \frac{Z(t, T) - Z(t, T + \alpha)}{\alpha Z(t, T + \alpha)}.$$

A ***libor-in-arrears*** or ***arrears FRA*** has identical cashflows and references the same libor, but the cashflows are made at time T ***not*** $T+\alpha$. Precisely the payout of a libor-in-arrears FRA at time T is

$$\alpha(L_T - K).$$

The ***forward libor-in-arrears rate*** $\widetilde{L}_t\,[T, T+\alpha]$, abbreviated to $\widetilde{L}_{tT}$, is defined to be the value of K that makes the arrears FRA have zero value at $t \le T$. We will see in the next section that in general

$$\widetilde{L}_{tT} \neq L_{tT}.$$

Heuristically, valuation of the libor-in-arrears FRA is complex because one can only deposit at libor from T until $T+\alpha$ (by definition). A contract that pays libor at time T cannot be easily replicated.

Note that there is no such thing as a libor-in-arrears 'spot rate' or 'fix'. Both the regular FRA and the libor-in-arrears FRA are derivatives of the (standard) libor rate.

14.2 Libor-in-arrears convexity correction

To tackle valuation of the arrears FRA, we compound its value up to time $T+\alpha$. Specifically, the arrears FRA has payout

$$D_K(T,T) = \alpha(L_T - K) \text{ at time } T,$$

and thus value

$$\alpha(L_T - K)(1+\alpha L_T) = \alpha(L_T - K) + \alpha^2(L_T^2 - KL_T) \text{ at time } T+\alpha.$$

Thus we can re-express

$$D_K(T,T) = \left(\alpha(L_T - K) + \alpha^2(L_T^2 - KL_T)\right) Z(T, T+\alpha).$$

Using the fundamental theorem we have

$$\frac{D_K(t,T)}{Z(t,T+\alpha)} = E_*\left(\alpha(L_T - K) + \alpha^2(L_T^2 - KL_T) \mid L_{tT}\right),$$

where E_* is the familiar risk-neutral expectation with respect to the forward numeraire $Z(t, T+\alpha)$. Therefore,

$$\begin{aligned} D_K(t,T) &= \alpha Z(t,T+\alpha)\left(E_*(L_T \mid L_{tT}) - K + \alpha(E_*(L_T^2 \mid L_{tT}) - KE_*(L_T \mid L_{tT}))\right) \\ &= \alpha Z(t,T+\alpha)(L_{tT} - K + \alpha(E_*(L_T^2 \mid L_{tT}) - KL_{tT})), \end{aligned}$$

since $E_*(L_T \mid L_{tT}) = L_{tT}$. Suppose we chose $K = L_{tT}$ (the naïve choice), then the value of the arrears FRA

$$D_K(t,T) = \alpha^2 Z(t,T+\alpha)(E_*(L_T^2 \mid L_{tT}) - L_{tT}^2) \geq 0,$$

by Jensen's inequality. Therefore, unless L_T is a constant, $\widetilde{L}_{tT}$ must be greater than L_{tT}. The value at time t of receiving the (random) libor L_T in arrears at T is higher than receiving the known forward libor L_{tT} at T. The difference $\widetilde{L}_{tT} - L_{tT}$ is known as the ***libor-in-arrears convexity correction***.

We can calculate this convexity correction using the result that calls (or caplets) are a spanning set. Using the identity from Chapter 11 Question 1 that

$$V(x) = V(0) + xV'(0) + \int_0^\infty V''(K)(x-K)^+ dK,$$

with $V(x) = \alpha x(1 + \alpha x)$, we have

$$\alpha L_T + \alpha^2 L_T^2 = \alpha L_T + \int_0^\infty 2\alpha^2 \left(L_T - K\right)^+ dK.$$

Result *Using the fundamental theorem with the $Z(t, T+\alpha)$ numeraire, we obtain the value at time t of a libor-in-arrears payment αL_T at time T is*

$$Z(t,T+\alpha)E_*(\alpha L_T(1+\alpha L_T)) = Z(t,T+\alpha)\alpha L_{tT} + \int_0^\infty 2\alpha C_K\left(t,T\right) dK,$$

a regular libor term, plus a positive convexity correction which is a function of caplet prices. In particular, receiving a libor-in-arrears payment will have positive vega.

In order to use a simple replication argument that we employed for the FRA, the floating libor rate must be received at the 'right' time $T + \alpha$, the maturity of the rate. In particular, three-month libor must be received after three months, or else there is a convexity adjustment. Receiving the floating rate and paying the fixed rate too soon, as in a libor-in-arrears FRA, is preferable in the sense that we would pay a higher fixed rate $\widetilde{L}_{tT}$ in return for receiving the floating rate early, than the L_{tT} for a regular FRA. This convexity correction is similar to that for a futures contract, where it is preferable to receive money earlier than the forward contract, if this occurs when interest rates are high.

The regular FRA can be replicated by a linear combination of ZCBs and thus can be priced from today's ZCB prices. The libor-in-arrears FRA is not linear in ZCB prices and to price it one needs to know the distribution of L_T.

Note The libor-in-arrears valuation includes a libor-squared term. Whilst now rare, libor-squared swaps received some publicity in the early 1990s when they were a component of the derivative contracts that caused significant losses at Gibson Greetings in 1994.

Table 14.1 Scenarios for libor fix.

Libor fix at T	Value of regular FRA	Value of arrears FRA	Total
4%	$+2\% \times 106 \div 1.04$	$-2\% \times 100$	>0
5%	$+1\% \times 106 \div 1.05$	$-1\% \times 100$	>0
6%	0	0	0
7%	$-1\% \times 106 \div 1.07$	$+1\% \times 100$	>0
8%	$-2\% \times 106 \div 1.08$	$+2\% \times 100$	>0

The intuition behind the libor-in-arrears convexity correction is illustrated well with a simple worked example. Suppose $\alpha = 1$ and the forward libor $L_{tT} = 6\%$. Suppose in addition that the forward libor-in-arrears $\widetilde{L}_{tT}$ also equals 6%. We will show that unless L_T is a constant, we can construct an arbitrage portfolio.

In particular, we construct the portfolio where we receive 6% and pay libor on \$106 million notional of a regular FRA, and we pay 6% and receive libor on \$100 million notional of a libor-in-arrears FRA. We can construct this portfolio at no cost.

We now calculate, in Table 14.1, the value of the portfolio at the fixing date T for different scenarios for the libor fix L_T.

Therefore, our portfolio is an arbitrage portfolio (unless L_T is a constant 6%), and we conclude empirically that the libor-in-arrears forward rate has to be higher than 6%. Note also that the benefit from receiving libor-in-arrears versus paying regular libor increases as the volatility of the libor rate increases, and thus we agree empirically that receiving libor-in-arrears has positive vega.

We summarize our findings in the following rule. Receiving a floating interest rate earlier than its contracted term is preferable, meaning we would pay higher than the regular forward rate to do so, and such a contract has positive vega. By contrast, receiving a rate later than its term is not-preferable, and has negative vega. Hull (2011) calls this the 'timing adjustment.'

14.3 Classic libor-in-arrears trade

For simplicity, assume that $\alpha = 1$ and so $T_{i+1} = T_i + 1$. Suppose $L_t[T_1, T_2] = 5.00\%$ and $L_t[T_2, T_3] = 5.20\%$. Then the following two FRAs both have zero value at time $t < T_1$.

(a) Receive $L_{T_1}[T_1, T_2]$ and pay 5.00% at T_2.

(b) Pay $L_{T_2}[T_2, T_3]$ and receive 5.20% at T_3.

Suppose Bank A believes that libor will increase by less than 20bp between T_1 and T_2, and so wants to enter both the trades above, which it can do at zero cost. This is known as a yield curve flattener trade. However, Bank B approaches Bank A with the following proposition. 'Doing the trade via two FRAs involves cashflows at different times. Why not simply combine the two zero-cost trades, then simplify the trade by having all the cashflows made at time T_2. Specifically, we suggest you receive $L_{T_1}[T_1, T_2]$, pay 5.00%, pay $L_{T_2}[T_2, T_3]$ and receive 5.20%, all at T_2. That is, net you receive $L_{T_1}[T_1, T_2]$ + 20bp and pay $L_{T_2}[T_2, T_3]$ at T_2.'

Bank B made the proposal even more attractive for Bank A. They claimed the trade suited their portfolio, so they could pay Bank A a spread of 21bp instead of 20bp. Bank A considered this a good deal and believed that they were making a profit of 1bp, in addition to implementing the trade they wanted. Therefore, Bank A executed the trade as described.

However, we can see upon closer inspection that, whilst Bank A believed the value of the trade to them was

$$L_t[T_1, T_2] - L_t[T_2, T_3] + 21\text{bp} = +1\text{bp},$$

the value of the trade at time $t < T_1$ was in fact

$$L_t[T_1, T_2] - \widetilde{L}_t[T_2, T_3] + 21\text{bp}.$$

The libor-in-arrears convexity correction $\widetilde{L}_t[T_2, T_3] - L_t[T_2, T_3]$ at t was in fact around 5bp. So in fact Bank A was losing (and Bank B making) 4bp on the trade.

This trade hid the subtlety of the libor-in-arrears convexity correction within a simple yield curve trade. The complexity was injected simply when the time of the payment of the second FRA was moved from T_3 to T_2.

14.4 Constant maturity swap contracts

We can extend arguments about the libor-in-arrears convexity correction to contracts where it is the swap rate, rather than the libor rate, that is paid early. These are known as ***constant maturity swap (CMS)*** contracts. Whilst libor-in-arrears payments are relatively easy to analyse, they now trade infrequently. CMS contracts, on the other hand, are widely traded in the market. A comprehensive analysis of CMS valuation is complex and beyond our scope, although key ideas transfer over directly from the libor-in-arrears example.

The simplest ***CMS contract*** is defined where one

$$\left.\begin{array}{ll}\text{pays} & K \\ \text{receives} & y_T[T, T_n]\end{array}\right\} \text{at a single time } T.$$

Since $y_T[T, T_n]$ is by definition the fixed rate such that the value of a swap from T to T_n has zero value, in a CMS contract we are receiving the swap rate earlier than its contracted term, that is, at time T instead of at payment dates T_i, for $i = 1, \ldots, n$. Thus, similar to the libor-in-arrears FRA, one would expect the ***CMS rate*** $\widetilde{y}_t[T, T_n]$, the value of K such that the CMS contract has zero value at time t, to be greater than $y_t[T, T_n]$, the forward swap rate.

Table 14.2 Swap rate scenarios.

Swap rate at T	**Value of swap**	**Value of CMS contract**	**Total**
5%	$+1\% \times P_T[T, T_n]$	$-1\% \times P_t[T, T_n]$	>0
6%	0	0	0
7%	$-1\% \times P_T[T, T_n]$	$+1\% \times P_t[T, T_n]$	>0

To show that this is indeed the case, we suppose that $\tilde{y}_t[T, T_n] = y_t[T, T_n] = 6\%$ and construct an arbitrage portfolio.

We can enter into the following two contracts at zero cost.

(a) A forward swap, where we receive 6% and pay libor from T to T_n.

(b) A CMS contract, where we pay 6% and receive $y_T[T, T_n]$ at time T.

We will execute (a) on a notional of 1 and (b) on a notional of $P_t[T, T_n]$ (which is a known quantity at t), and consider the value of this portfolio at time T. Note that the pv01 $P_T[T, T_n]$ of the swap at T will differ depending on the level of interest rates. In particular, when the swap rate is 7%, the pv01 will be lower than $P_t[T, T_n]$—the pv01 when the rate was at 6%. Similarly, the pv01 when the rate is 5% is higher. Hence we obtain the portfolio values for the scenarios shown in Table 14.2.

Thus we have constructed an arbitrage portfolio, and so we conclude that $\tilde{y}_t[T, T_n] > y_t[T, T_n]$. The difference $\tilde{y}_t[T, T_n] - y_t[T, T_n]$ is called the ***CMS convexity correction***, and depends on the distribution of $y_T[T, T_n]$. It is generally increasing in T, since for larger T the swap rate typically has higher variance, and increasing in $(T_n - T)$, since there is a larger timing difference. The CMS correction can be significant, for example over 50bp for longer dated expiries on long-dated rates.

Again we see that receiving a rate early (the T to T_n swap rate at time T, instead of over the 'right' period T to T_n) is preferable, in the sense that one would pay a higher fixed rate to receive it early than over its natural time. A position where one receives the CMS rate has positive vega.

14.5 EXERCISES

1. **Interest rate vega**
   For each of the following positions (a)-(g) in Table 14.3, determine whether Bank A has vega ≥ 0, vega ≤ 0, there is no volatility exposure, or the volatility exposure is indeterminate ('?').

Table 14.3 Interest rate vega.

	Bank A position	Vega: ≥ 0, ≤ 0, $= 0$ or ?
(a)	Bank A receives 10-year CMS q, pays 3mL q, for 5 years	
(b)	Bank A pays 3mL q, receives 3mL in arrears q, for 10 years	
(c)	Bank A receives 7% m, pays 6mL s/a, for 7 years	
(d)	Bank A pays 1mL m, receives 6mL s/a, for 5 years	
(e)	Bank A pays 1mL s/a, receives 6mL s/a, for 5 years	
(f)	Bank A is long a futures contract on a fixed rate bond	
(g)	Bank A is long a forward contract on a fixed rate bond	

2. **Digital caplet in arrears**
   Consider an option that pays

$$\alpha \text{ at time } T \text{ if } L_T > K, \text{ and zero otherwise.}$$

By considering the payout of this option (a 'digital caplet in arrears') at $T + \alpha$, show that its price $\tilde{D}_K(t, T)$ at time $t \leq T$ is given by

$$\tilde{D}_K(t, T) = (1 + \alpha K)\, D_K(t, T) + \alpha C_K(t, T),$$

where $D_K(t, T)$ is the price of a regular digital caplet, and $C_K(t, T)$ is the price of a K-strike caplet with maturity T.

15

The Brace–Gatarek–Musiela framework

We give a brief introductory sketch of the Brace–Gatarek–Musiela (BGM) model, which provides a framework for the consistent valuation of a range of interest rate options. We describe the BGM volatility surface and explore how the prices of various interest rate derivatives depend on this surface.

15.1 BGM volatility surface

For a fixed T, write $X_t = \log L_{tT}$. Then under the lognormal risk-neutral distribution we have

$$X_T \mid X_t = X_t - \frac{1}{2}\sigma^2(T-t) + \sigma\sqrt{T-t}\,W, \; W \sim N(0,1),$$

where σ is the volatility of $X_T \mid X_t$, and we are using the numeraire $Z(t, T+\alpha)$. Libor rates with different maturities T will in general have different volatilities, so strictly our notation should include the dependence $\sigma = \sigma_T$. Note also we would need a different numeraire $Z(t, T+\alpha)$ for each forward libor in order to use the pricing machinery developed in Chapter 12.

Suppose we divide $[t, T] = [t_0, \; t_n]$ into intervals $[t_i, \; t_{i+1}]$, $i = 0, \ldots, n-1$, with piecewise constant volatility $\sigma_{i,T}$ on each interval. Then

$$X_{t_n} - X_{t_0} = (X_{t_n} - X_{t_{n-1}}) + \cdots + (X_{t_1} - X_{t_0})$$
$$\text{with } X_{t_{i+1}} - X_{t_i} \sim N\left(-\frac{1}{2}\sigma_{i,T}^2(t_{i+1}-t_i),\; \sigma_{i,T}^2(t_{i+1}-t_i)\right).$$

Since the sum of independent normals is itself normal, $X_{t_n} - X_{t_0}$ is normal with

$$\text{Var}\left(X_{t_n} - X_{t_0}\right) = \sum_{i=0}^{n-1} \sigma_{i,T}^2(t_{i+1}-t_i).$$

Since every continuous function can be uniformly approximated by piecewise linear functions, then for a positive function $\sigma(t, T)$ of t and T (the continuous limit of the piecewise $\sigma_{i,T}$), we have

$$\mathrm{Var}(\log L_T - \log L_{t_0 T}) = \int_{t_0}^{t_n} \sigma^2(t, T)dt.$$

Thus, the risk-neutral distribution for L_T under the forward numeraire $Z(t, T + \alpha)$ is given by

$$L_T \mid L_{tT} \sim \mathrm{lognormal}\left(\log L_{tT} - \frac{1}{2}\sigma^2(T - t), \sigma^2(T - t)\right),$$

where

$$\sigma^2(T - t) = \int_t^T \sigma^2(u, T)du.$$

For fixed T, the function $\sigma(t, T), 0 \leq t \leq T$ represents the instantaneous volatility from t to $t + \Delta t$ of the forward libor L_{tT}. The surface $\{\sigma(t, T), 0 \leq t \leq T \leq \infty\}$ is known as the ***BGM volatility surface***, after the Brace–Gatarek–Musiela (1997) model, which first explicitly described the evolution of libor rates in terms of their individual volatilities.

15.2 Option price dependence on BGM volatility surface

The price of a caplet on $L_{T_i}[T_i, T_{i+1}]$, where $T_{i+1} = T_i + \alpha$, depends on the volatilities

$$\{\sigma(t, T_i); 0 \leq t \leq T_i\}.$$

A T_i by T_j cap price depends on the volatilities in a trapezium

$$\{\sigma(t, T); T_i \leq T \leq T_{j-1}, 0 \leq t \leq T\},$$

as shown in Figure 15.1. The last caplet is on the libor rate $L_{T_{j-1}}[T_{j-1}, T_j]$ and expires at T_{j-1}.

We now consider the dependence of the swaption price on the BGM volatility surface. Recall that in Chapter 12 we priced a swaption using the swap numeraire. We let the risk-neutral distribution of the swap rate $y_{T_0}[T_0, T_n]$ be lognormal with expected value $y_t[T_0, T_n]$ and volatility σ^* (where we use the temporary notation σ^* to distinguish swaption volatility from caplet volatility).

How does one relate the volatilities $\sigma(t, T)$ of the libor forwards with the swaption volatility σ^*? Recall that

$$y_t[T_0, T_n] = \frac{\sum_{i=1}^n L_t[T_{i-1}, T_i]\alpha_i Z(t, T_i)}{\sum_{i=1}^n \alpha_i Z(t, T_i)},$$

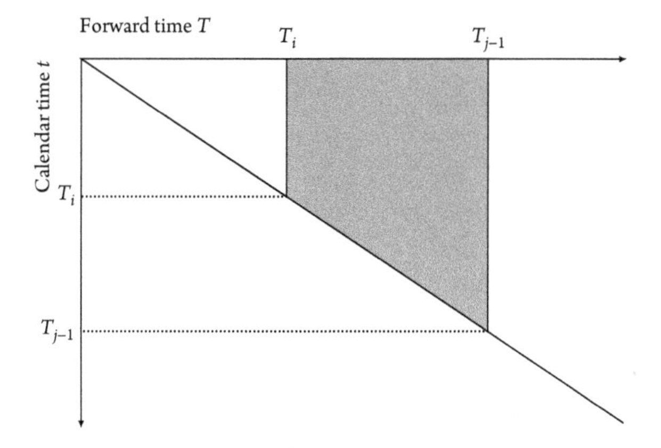

Figure 15.1 BGM volatility surface exposure of cap

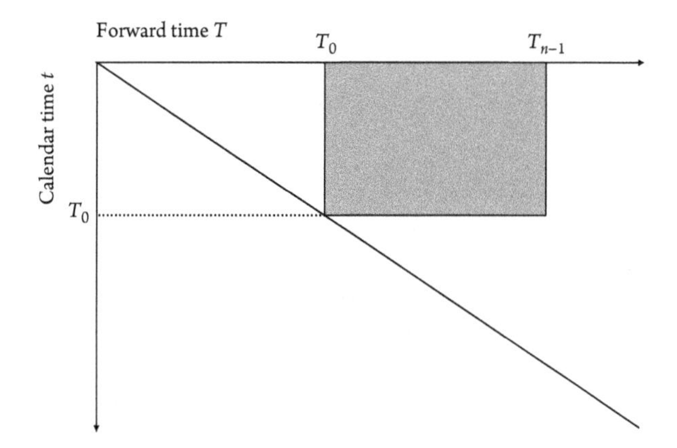

Figure 15.2 BGM volatility surface exposure of swaption

a weighted average of libor forwards. The $Z(t, T_i)$ themselves depend on the libor forward rates. However, were we to approximate the ZCB prices as constant (given their dependence on interest rates is of lower order than the rates themselves), heuristically we can see that the variance of $y_{T_0}[T_0, T_n]$ will depend on the volatilities of $L_t[T_{i-1}, T_i]$ for $0 \leq t \leq T_0$, and the correlation between them. In particular, the volatility, and thus the price, of a swaption depends on a rectangle of the volatility surface $\{\sigma(t, T); 0 \leq t \leq T_0, T_0 \leq T \leq T_{n-1}\}$, as shown in Figure 15.2.

Whilst we do not explore correlation structure in detail here, note that the swaption price will also depend upon the correlations $\rho(t, T_i, T_j)$ between $L_t[T_i, T_{i+1}]$ and $L_t[T_j, T_{j+1}]$, for $0 \leq t \leq T_0$, $1 \leq i < j \leq n-1$. In particular, if the $\sigma(t, T)$ (and hence cap prices) remain constant, then when correlation increases, swaption volatilities and prices increase.

Volatility trading in practice involves understanding the dependences of particular derivative products on the BGM volatility surface. Different derivatives can have inter-related

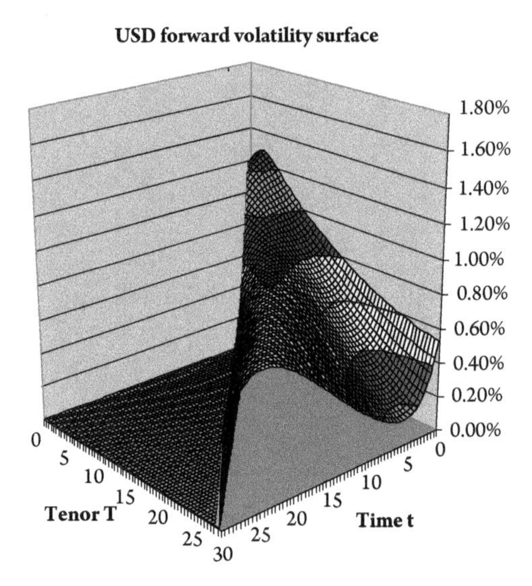

Figure 15.3 Typical parametric BGM volatility surface

dependences, which allow assessments of relative value. For example, a ***reset cap*** is a series of reset caplets, each with payout equal to $\alpha \max\{L_{T_i}[T_i, T_{i+1}] - L_{T_{i-1}}[T_{i-1}, T_i], 0\}$ at time T_i (in other words, where the strike of the caplet is the previous libor fix). The value of the reset caplet at time 0 will depend on the volatilities $\{\sigma(t, T_i), 0 \le t \le T_i\}$ and $\{\sigma(t, T_{i-1}), 0 \le t \le T_{i-1}\}$. However, for the period $0 \le t \le T_{i-1}$ the reset caplet is a derivative of the difference between the two libor rates, and thus has correlation exposure. From T_{i-1} to T_i, the reset caplet is an option just on $L_{T_i}[T_i, T_{i+1}]$ as the strike has fixed, and does not have correlation exposure. A reset cap thus has BGM exposure with similarities both to caps and swaptions.

A typical implementation of BGM often involves assuming a relatively simple function for the correlation structure, then proposing reasonable functional forms for $\sigma(t, T)$. A non-linear calibration follows to obtain the best parametric fit of $\sigma(t, T)$ to market option prices. This smooth surface provides rich-cheap indicators for options, by comparing the price given by the BGM volatility surface to actual market prices. One can also discretize the smooth surface then perturb it to obtain an exact nonparametric fit to market prices. Notable features of the nonparametric fit can then be analysed. Typical parametric and nonparametric fits to the US dollar options market are shown in Figures 15.3 and 15.4 respectively. For more details about the interpretation of the BGM volatility surface and its use in trading options, see Blyth (2004).

Note that when we priced an individual caplet, we were able to set $E_*(L_T|L_{tT}) = L_{tT}$ because we used the forward numeraire $Z(t, T + \alpha)$. This numeraire is specific to the libor rate fixing at T, which is itself a ratio of an asset to this numeraire, and hence a martingale. Similarly, the swap rate is a martingale under the swap numeraire. How does one reconcile these different numeraires in the BGM model, where one is attempting to evolve

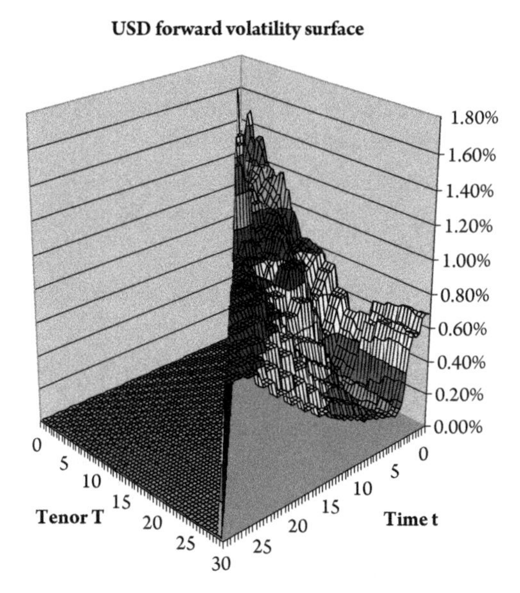

Figure 15.4 Typical nonparametric BGM volatility surface

many libor rates simultaneously under a consistent risk-neutral distribution? Given that one can only choose one numeraire, inevitably not all libor rates and swap rates will be martingales. Implementation of BGM typically involves using the money market numeraire, and requires calculation of the expected value for each libor rate, a procedure that is beyond our scope. Importantly, however, the volatility structure is not affected by the change of numeraire (similarly to the result we obtained in Chapter 10, where the volatility of the limiting lognormal distribution was the same under both actual and risk-neutral probabilities). Insights we obtain from analysing the volatility surface translate to the money market numeraire from the individual forward numeraires.

15.3 EXERCISES

1. **BGM surfaces I**

   Suppose the current date is April 2011. On the triangular BGM volatility surface $\sigma\,(t, T)$, $0 \leq t \leq T$, highlight the areas of volatility upon which depend the prices of the following interest rate options.

   (a) A 1yr-into-2yr European swaption (that is, April 2012 option into a swap with start date April 2012 and end date April 2014).

   (b) A 1yr-into-3yr European swaption.

(c) A 2yr-into-1yr European swaption.

(d) A 1yr-by-3yr cap.

(e) A one-year option into a swap with start date April 2013 and end date April 2014. (This is called a ***midcurve swaption***.)

(f) A caplet on three-month libor L_T, with $T = 16$ September 2013.

(g) A call option with exercise date 16 September 2013 on the underlying EDU3, the September 2013 Eurodollar contract.

(h) A ***midcurve*** call option on EDU3, that is, an option with exercise date 15 September 2011 on EDU3.

(i) A '1yr-into-1yr-1yr' ***forward-starting swaption***. The strike of the swaption is determined on April 2012 as the then forward swap rate (for a swap with dates April 2013 to April 2014), and the exercise date of the swaption is April 2013, into the swap with start date April 2013 and end date April 2014. (Derivatives of this form trade reasonably frequently in the US, and are sometimes referred to as '***forward volatility***' trades.)

2. **BGM surfaces II**

On the triangular BGM volatility surface $\sigma(t, T)$, $0 \leq t \leq T$, highlight the areas upon which depend the prices at time $t = 0$ of the following interest rate derivatives. Assume $0 < T_1 < T_2$.

(a) A 2yr-into-5yr European swaption.

(b) A 2yr-by-7yr cap on three-month libor.

(c) A 2yr-into-5yr Bermudan swaption with quarterly exercises.

(d) A contract that pays

$$\max\{(L_{T_1}[T_2, T_2 + \alpha] - L_{T_1}[T_1, T_1 + \alpha]), 0\} \text{ at time } (T_1 + \alpha).$$

(e) A contract that pays

$$\max\{(L_0[T_2, T_2 + \alpha] - L_0[T_1, T_1 + \alpha]), 0\} \text{ at time } (T_1 + \alpha).$$

(f) A contract (known as a ***reset cap***) that pays

$$\max\{(L_{T_2}[T_2, T_2 + \alpha] - K), 0\} \text{ at time } (T_2 + \alpha), \text{ where } K = L_{T_1}[T_1, T_1 + \alpha].$$

(g) A contract that pays

$$(L_{T_2}[T_2, T_2 + \alpha] \text{ at time } (T_2 + \alpha)) - (L_{T_1}[T_1, T_1 + \alpha] \text{ at time } (T_1 + \alpha)).$$

PART V

Towards Continuous Time

16

Rough guide to continuous time

In Part III we developed replication, no-arbitrage and the concept of risk-neutral pricing on a binomial tree, with two possible states of the world at each step. We showed that no-arbitrage was equivalent to the ratios of asset prices to a numeraire being martingales under the risk-neutral distribution. Although finance is essentially performed in discrete time, it is useful to develop continuous-time models, which can be easier to manipulate. In Chapter 10 we simply took the limit of the tree for a fixed T and calculated expected values using the limiting continuous distribution for S_T, and assumed without proof that the results from discrete time, in particular the fundamental theorem, transferred directly to the continuous case. In general, mathematics of continuous-time finance is non-trivial and beyond our scope. This chapter is designed to give a flavour of continuous-time theory by briefly presenting some key concepts concerning stochastic processes. In particular, we introduce Brownian motion as the limit of the binomial tree, and show how we can use Ito's lemma to tackle the change of numeraire and risk-neutral distributions in continuous time, arriving elegantly at the Black–Scholes formula. For a more extensive exposition of continuous-time theory, see Shreve (2004) or Joshi (2008).

16.1 Brownian motion as random walk limit

$$\text{Let } \xi_i \text{ be IID and equal to } \begin{cases} +1 & \text{with probability } 1/2 \\ -1 & \text{with probability } 1/2 \end{cases}$$

and consider the symmetric random walk $\sum_{i=1}^{N} \xi_i$. By the central limit theorem

$$W_N = \sum_{i=1}^{N} \frac{1}{\sqrt{N}} \xi_i \to W \text{ as } N \to \infty, \text{ where } W \sim N(0, 1).$$

For fixed $t > 0$ define

$$W_{N,t} = \sum_{i=1}^{[Nt]} \frac{1}{\sqrt{N}} \xi_i = \sqrt{t} \sum_{i=1}^{[Nt]} \frac{1}{\sqrt{Nt}} \xi_i,$$

where $[Nt]$ is the closest integer to Nt. Then

$$W_{N,t} \to W_t \sim N(0, t) \text{ as } N \to \infty.$$

It is possible to extend this result and take the limit for all t simultaneously, although this theory is beyond our scope. The limit W_t, the continuous-time limit of a symmetric random walk, is called ***Brownian motion*** or a Wiener process. It has the following properties.

(a) $W_0 = 0$.
(b) For all $t \leq s$, $W_s - W_t \sim N(0, s - t)$.
(c) $W_s - W_t$ is independent of W_t, $t \leq s$.

The intuitive understanding of Brownian motion as the limit of the symmetric random walk with ever smaller time steps is helpful. Equivalently, it can be thought of as a process with independent normal increments, since for any t and $\Delta t > 0$

$$W_{t+\Delta t} - W_t \sim N(0, \Delta t).$$

Note that

$$E(W_s \mid W_t) = E((W_s - W_t) + W_t \mid W_t) = E(W_s - W_t \mid W_t) + W_t = W_t$$

so Brownian motion is a martingale.

16.2 Stochastic differential equations and geometric Brownian motion

For $\Delta t > 0$ we can write rigorously a difference equation such as

$$S_{t+\Delta t} - S_t = \mu \Delta t + \sigma (W_{t+\Delta t} - W_t) = \mu \Delta t + \sigma \sqrt{\Delta t} W, \text{ where } W \sim N(0, 1).$$

We write the limit as $\Delta t \to 0$

$$dS_t = \mu dt + \sigma dW_t,$$

or, for more general functions $\mu(S_t, t)$ and $\sigma(S_t, t)$,

$$dS_t = \mu(S_t, t)dt + \sigma(S_t, t)dW_t.$$

This is a stochastic differential equation, the general theory of which we do not tackle here. The meaning of dW_t is non-trivial, although the discrete analogue $\sqrt{\Delta t}\, W$, where $W \sim N(0,1)$, is helpful for conceptual understanding.

For constant μ and σ the solution of the stochastic differential equation is

$$S_T = \mu T + \sigma\sqrt{T}W, \text{ where } W \sim N(0,1),$$

and for piecewise constant functions we have

$$S_T = \sum_{i=1}^{n} \mu_i \left(t_{i+1} - t_i\right) + \left(\sqrt{\sum_{i=1}^{n} \sigma_i^2 \left(t_{i+1} - t_i\right)}\right) W.$$

What are possible functions $\mu(S_t, t)$ and $\sigma(S_t, t)$ to model the behaviour of a stock price? A reasonable assumption is that the drift μ of a stock is proportional to its price, since a stock split should not change the dynamics of the stock price, so we propose $\mu\,(S_t, t) = \mu S_t$. With no random term, the ordinary differential equation would be

$$dS_t = \mu S_t dt \Rightarrow S_t = S_0 e^{\mu t}.$$

For $\sigma(S_t, t)$, one might assume that uncertainty about the stock's percentage return (rather than uncertainty about its absolute price) is constant. That is, the probability of a stock move from 100 to outside $(99, 101)$ equals the probability of a move from 50 to outside $(49.5, 50.5)$. Equivalently, over a small interval Δt

$$\operatorname{Var}\left(\frac{\Delta S_t}{S_t}\right) \approx \sigma^2 \Delta t \Rightarrow \operatorname{Var}\left(\Delta S_t\right) \approx \sigma^2 S_t^2 \Delta t.$$

Therefore, one possible process for the evolution of the stock is

$$dS_t = \mu S_t dt + \sigma S_t dW_t,$$

which is the definition that the stock follows ***geometric Brownian motion***. Its discrete-time analogue is

$$\frac{\Delta S_t}{S_t} = \mu\Delta t + \sigma\sqrt{\Delta t}W, \text{ where } W \sim N\left(0,1\right).$$

What is the distribution of S_T, assuming geometric Brownian motion? We need additional tools to address this question.

16.3 Ito's lemma

Consider a function $f(X_t, t)$ of X_t, ignoring temporarily that X_t may be random. Then a Taylor series gives

$$\Delta f = \frac{\partial f}{\partial t}\Delta t + \frac{\partial f}{\partial X_t}\Delta X_t + \frac{1}{2}\frac{\partial^2 f}{\partial X_t^2}(\Delta X_t)^2 + \frac{\partial^2 f}{\partial X_t \partial t}\Delta X_t \Delta_t + \frac{1}{2}\frac{\partial^2 f}{\partial t^2}(\Delta t)^2 + \ldots$$

For non-random X_t

$$df = \frac{\partial f}{\partial t}dt + \frac{\partial f}{\partial X_t}dX_t$$

by the chain rule for differentiation. Now suppose that

$$\Delta X_t = \mu \Delta t + \sigma\sqrt{\Delta t}W, \text{ where } W \sim N(0,1),$$

and so

$$(\Delta X_t)^2 = \sigma^2 \Delta t W^2 + O\left(\Delta t^{\frac{3}{2}}\right).$$

Since $E(W^2) = 1$ and $\mathrm{Var}(W^2) = 2$ then $(\Delta X_t)^2 \to \sigma^2 \Delta t$ as $\Delta t \to 0$, and we have

$$\Delta f \approx \frac{\partial f}{\partial t}\Delta t + \frac{\partial f}{\partial X_t}\Delta X_t + \frac{1}{2}\frac{\partial^2 f}{\partial X_t^2}\sigma^2 \Delta t.$$

Taking limits we obtain ***Ito's lemma.***

If $dX_t = \mu(X_t, t)dt + \sigma(X_t, t)dW_t$, then

$$\begin{aligned} df &= \frac{\partial f}{\partial t}dt + \frac{\partial f}{\partial X_t}dX_t + \frac{1}{2}\frac{\partial^2 f}{\partial X_t^2}\sigma^2 dt \\ &= \left(\frac{\partial f}{\partial t} + \frac{\partial f}{\partial X_t}\mu(X_t, t) + \frac{1}{2}\frac{\partial^2 f}{\partial X_t^2}\sigma^2(X_t, t)\right)dt + \frac{\partial f}{\partial X_t}\sigma(X_t, t)\,dW_t. \end{aligned}$$

Ito's lemma is easiest to remember in the following form. If $dx = \mu dt + \sigma dW_t$, then

$$df = \frac{\partial f}{\partial t}dt + \frac{\partial f}{\partial x}dx + \frac{1}{2}\frac{\partial^2 f}{\partial x^2}dx^2,$$

where dx^2 is defined by the identities $dt^2 = 0$, $dtdW_t = 0$ and $(dW_t)^2 = dt$.

We now apply Ito's lemma to $\log S_t$, where S_t follows geometric Brownian motion. Let $f(S_t) = \log S_t$ and so

$$\frac{\partial f}{\partial t} = 0, \quad \frac{\partial f}{\partial S_t} = \frac{1}{S_t} \text{ and } \frac{\partial^2 f}{\partial S_t^2} = -\frac{1}{S_t^2}.$$

Ito's lemma gives

$$d(\log S_t) = \left(\frac{1}{S_t}\mu S_t - \frac{1}{2}\sigma^2 S_t^2 \frac{1}{S_t^2}\right) dt + \sigma S_t \frac{1}{S_t} dW_t = \left(\mu - \frac{1}{2}\sigma^2\right) dt + \sigma dW_t.$$

Therefore, $\log S_t$ follows standard Brownian motion and is normally distributed. Specifically,

$$\log S_T \mid S_t \sim N\left(\log S_t + \left(\mu - \frac{1}{2}\sigma^2\right)(T-t), \sigma^2(T-t)\right),$$

and we have shown that under geometric Brownian motion the distribution of $S_T \mid S_t$ is lognormal, equal to the limiting distribution of the binomial tree in Chapter 10.

16.4 Black–Scholes equation

Now consider the price of a call (or other derivative contract) $C_K(t, T)$, which is a function of the stock price. If the stock follows geometric Brownian motion, then by Ito's lemma the option price satisfies the stochastic differential equation

$$dC_K(t,T) = \left(\frac{\partial C_K(t,T)}{\partial t} + \frac{\partial C_K(t,T)}{\partial S_t}\mu S_t + \frac{1}{2}\frac{\partial^2 C_K(t,T)}{\partial S_t^2}\sigma^2 S_t^2\right) dt + \frac{\partial C_K(t,T)}{\partial S_t}\sigma S_t dW_t.$$

Suppose we construct at t a portfolio Π consisting of long one option and short $\frac{\partial C_K(t,T)}{\partial S_t}$ of stock. Its price Π_t at time t satisfies

$$\Pi_t = C_K(t,T) - \frac{\partial C_K(t,T)}{\partial S_t} S_t.$$

Using the expressions for $dC_K(t, T)$ and dS_t, we obtain

$$d\Pi_t = \left(\frac{\partial}{\partial t}C_K(t,T) + \frac{1}{2}\frac{\partial^2 C_K(t,T)}{\partial S_t^2}\sigma^2 S_t^2\right) dt.$$

This portfolio instantaneously at t has no exposure to the term dW_t. Therefore, it is instantaneously a replicating portfolio for the money market account and must grow at rate r. So

$$d\Pi_t = r\Pi_t dt \Rightarrow \frac{\partial C_K(t,T)}{\partial t} + \frac{1}{2}\frac{\partial^2 C_K(t,T)}{\partial S_t^2}\sigma^2 S_t^2 = r\left(C_K(t,T) - \frac{\partial C_K(t,T)}{\partial S_t} S_t\right).$$

Therefore, we obtain

$$\frac{\partial C_K(t,T)}{\partial t} + r\frac{\partial C_K(t,T)}{\partial S_t} S_t - rC_K(t,T) + \frac{1}{2}\frac{\partial^2 C_K(t,T)}{\partial S_t^2}\sigma^2 S_t^2 = 0,$$

the ***Black–Scholes partial differential equation*** for a European derivative contract. The solution to this partial differential equation, under the boundary conditions $C_K(T,T) = (S_T - K)^+$, is the Black–Scholes formula we derived probabilistically in Chapter 10. We thus have a new approach for option pricing via partial differential equations. For further details see, for example, Wilmott (1998).

Compare this approach to how we replicated the ZCB on a binomial tree with an option and a particular holding Δ of stock. Here in continuous time, the analogue to Δ is given by the instantaneous hedge $\frac{\partial C_K(t,T)}{\partial S_t}$, the delta of the option.

16.5 Ito and change of numeraire

For two processes X_t, Y_t such that

$$dX_t = \mu\,(X_t,t)\,dt + \sigma\,(X_t,t)\,dW_t$$
$$dY_t = \nu\,(Y_t,t)\,dt + \tau\,(Y_t,t)\,dW_t$$

then $f\,(X_t, Y_t, t)$ satisfies the stochastic differential equation

$$df = \frac{\partial f}{\partial t}dt + \frac{\partial f}{\partial X_t}dX_t + \frac{\partial f}{\partial Y_t}dY_t + \frac{1}{2}\frac{\partial^2 f}{\partial X_t^2}dX_t^2 + \frac{\partial^2 f}{\partial X_t \partial Y_t}dX_t dY_t + \frac{1}{2}\frac{\partial^2 f}{\partial Y_t^2}dY_t^2,$$

with higher order terms again defined by identities $(dW_t^2) = dt,\ dt^2 = 0$ and $dW_t dt = 0$. Setting $f\,(x,y) = xy$ we obtain

$$d\,(X_t Y_t) = X_t dY_t + Y_t dX_t + dX_t dY_t.$$

This result is useful when considering the ratio of an asset to a numeraire. In particular, suppose again that S_t follows geometric Brownian motion $dS_t = \mu S_t dt + \sigma S_t dW_t$. The money market account $M_t = e^{rt}$ satisfies $dM_t = rM_t dt$, and so

$$d\left(\frac{1}{M_t}\right) = -\frac{1}{M_t^2}dM_t = -\frac{r}{M_t}dt.$$

We now apply Ito's lemma to the ratio S_t/M_t. Since there is no dW_t term in the differential equation for dM_t,

$$\begin{aligned} d\left(\frac{S_t}{M_t}\right) &= \frac{1}{M_t}dS_t + S_t d\left(\frac{1}{M_t}\right) \\ &= \frac{1}{M_t}\left(\mu S_t dt + \sigma S_t dW_t\right) + S_t\left(-\frac{r}{M_t}\right)dt = (\mu - r)\frac{S_t}{M_t}dt + \sigma\frac{S_t}{M_t}dW_t. \end{aligned}$$

Thus S_t/M_t follows geometric Brownian motion and we have

$$\log\left(\frac{S_T}{M_T} \mid S_t\right) \sim N\left(\log\left(\frac{S_t}{M_t}\right) + (\mu - r)(T-t) - \frac{1}{2}\sigma^2(T-t), \sigma\sqrt{T-t}\right).$$

In order for S_t/M_t to be a martingale, we must have $\mu = r$, that is, the drift term is zero.

Letting $f(S_t) = F(t,T) = S_t e^{r(T-t)}$, the forward price at time t, we have

$$\frac{\partial f}{\partial t} = -rS_t e^{r(T-t)} = -rF(t,T), \text{ and } \frac{\partial f}{\partial S_t} = e^{r(T-t)}.$$

So Ito's lemma gives the following stochastic differential equation for the forward price,

$$dF(t,T) = (\mu - r)F(t,T)dt + \sigma F(t,T)dW_t.$$

In the risk-neutral world with respect to the money market account we have $\mu = r$, and the stochastic differential equation for the forward price simply becomes $dF(t,T) = \sigma F(t,T)dW_t$. Therefore,

$$\log F(T^*,T) \mid F(t,T) \sim N\left(\log F(t,T) - \frac{1}{2}\sigma^2(T^*-t), \sigma^2(T^*-t)\right)$$
$$\text{and } E\left(F(T^*,T) \mid F(t,T)\right) = F(t,T)$$

for $t \leq T^* \leq T$. We have therefore shown that the forward price is a martingale under the risk-neutral distribution. We know from the fundamental theorem that we must have this result, since the forward price is itself the ratio of the stock to the ZCB.

We can begin to see the formation of a continuous-time analogue of our work on numeraires, martingales and the fundamental theorem. First we choose an appropriate numeraire, often chosen to simplify calculation. Given the stochastic differential equation for the underlying asset, we calculate using Ito's lemma the stochastic differential equation for assets rebased by that numeraire. By the fundamental theorem, the absence of arbitrage is equivalent to the rebased assets being martingales, which is itself equivalent to the stochastic differential equation having zero drift. We can thus determine the risk-neutral distribution with respect to the numeraire by imposing this condition.

As a further example, suppose we adopt the stock as numeraire, and consider the ratio M_t/S_t. By Ito's lemma,

$$d\left(\frac{M_t}{S_t}\right) = \frac{1}{S_t}dM_t + d\left(\frac{1}{S_t}\right)M_t.$$

Again, by Ito's lemma,

$$d\left(\frac{1}{S_t}\right) = -\frac{1}{S_t^2}\left(\mu S_t dt + \sigma S_t dW_t\right) + \frac{1}{S_t^3}\sigma^2 S_t^2 dt = \left(\sigma^2 - \mu\right)\frac{1}{S_t}dt - \frac{\sigma}{S_t}dW_t.$$

Therefore,

$$d\left(\frac{M_t}{S_t}\right) = \frac{M_t}{S_t}\left(r - \mu + \sigma^2\right) dt - \sigma \frac{M_t}{S_t} dW_t.$$

So M_t/S_t is a martingale $\Longleftrightarrow \mu = r + \sigma^2$, and the risk-neutral distribution under the stock numeraire is given by setting $\mu = r + \sigma^2$. Under this condition, S_t satisfies

$$dS_t = \left(r + \sigma^2\right) S_t dt + \sigma S_t dW_t.$$

The change of numeraire machinery is a powerful tool in finance, and we immediately give one application here to the Black–Scholes formula.

We can split the call payout $(S_T - K)^+ = (S_T - K)\, I\{S_T \geq K\}$ into two parts: **A** $S_T I\{S_T \geq K\}$; and **B** $-KI\{S_T \geq K\}$. To price **A**, consider the contract with price at time t denoted by $D^A(t,T)$, and payout at time T $D^A(T,T) = S_T I\{S_T \geq K\}$. By the fundamental theorem,

$$\frac{D^A(t,T)}{S_t} = E_{**}\left(\frac{D^A(T,T)}{S_T} \mid S_t\right) = P^{**}\left(S_T \geq K \mid S_t\right),$$

the probability P^{**} being the risk-neutral distribution with respect to the stock numeraire. Therefore,

$$D^A(t,T) = S_t P^{**}\left(S_T \geq K \mid S_t\right).$$

Under the risk-neutral distribution with respect to the stock numeraire, we set $\mu = r + \sigma^2$ and

$$\log S_T \mid S_t \sim N\left(\log S_t + \left(r + \frac{1}{2}\sigma^2\right)(T-t),\ \sigma^2(T-t)\right).$$

Therefore,

$$P^{**}\left(S_T \geq K \mid S_t\right) = 1 - \Phi\left(\frac{\log K - \log S_t - \left(r + \frac{1}{2}\sigma^2\right)(T-t)}{\sigma\sqrt{T-t}}\right)$$

$$= \Phi\left(\frac{\log\left(\frac{S_t}{K}\right) + \left(r + \frac{1}{2}\sigma^2\right)(T-t)}{\sigma\sqrt{T-t}}\right).$$

For **B**, we use the money market account as the numeraire, and consider the derivative with payout at time T $D^B(T,T) = KI\{S_T \geq K\}$. By the fundamental theorem,

$$\frac{D^B(t,T)}{M_t} = E_*\left(\frac{D^B(T,T)}{M_T}\right) = Ke^{-rT} P^*\left(S_T \geq K \mid S_t\right).$$

Under P^*, the risk-neutral distribution with respect to the money market account, we set $\mu = r$ and so

$$\log S_T \mid S_t \sim N\left(\log S_t + \left(r - \frac{1}{2}\sigma^2\right)(T - t),\ \sigma^2(T - t)\right).$$

Combining our results for **A** and **B** we obtain—without calculating any tail integrals or solving partial differential equations—the Black–Scholes formula, which seems an apt place to end.

GLOSSARY OF NOTATION

Symbol	Definition	Chapter
$B_c^{FXD}(t)$	Price at t of fixed rate bond with coupon c	4
$B^{FL}(t)$	Price at t of floating rate bond	4
$B_K(t, T_0, T_n)$	Price at t of K-strike Bermudan payer swaption with first exercise date T_0 on swap ending at T_n	13
$B_{K,\lambda}(t, T)$	Price at t of call butterfly with centre K and half-width $1/\lambda$	11
c	Bond coupon	1
$C_K(t, T)$	Price at t of call (or caplet) with strike K and exercise date T	7, 12
$\tilde{C}_K(t, T)$	Price at t of American call	7
$C^{F_K}(t, T)$	Price at t of call on forward contract	7
$C_K(t, T_0, T_n)$	Price at t of T_0 by T_n cap with strike K	13
$D(t, T)$	Price at t of derivative contract with maturity T	9
$D^A(t, T)$	Price at t of derivative contract **A** with maturity T	16
$D_K(t, T)$	Price at t of digital call with strike K and maturity T	11
$f(x)$	Probability density function	9
f_{12}	Forward interest rate from T_1 to T_2	3
$F(t, T)$	Forward price at t for forward contract with maturity T	2
$F(t, T_1, T_2)$	Forward price at t for forward contract with maturity T_1 on zero coupon bond with maturity T_2	3
$g(S_T)$	Derivative payout at maturity T	2
K	Delivery price, strike price or fixed rate	2, 4, 7
$L_T[T, T+\alpha], L_T$	Libor rate for period $[T, T+\alpha]$	3
$L_t[T, T+\alpha], L_{tT}$	Forward libor rate at t for period $[T, T+\alpha]$	3
$\widetilde{L}_t[T, T+\alpha], \widetilde{L}_{tT}$	Forward libor-in-arrears rate at t for period $[T, T+\alpha]$	14

Symbol	Definition	Chapter
M_t	Value at t of money market account	1
N_t	Price at t of numeraire	9
p, P	Probability, probability distribution	8
p^*, P^*	Risk-neutral probability, probability distribution	8, 9
$P_K(t, T)$	Price at t of put with strike K and exercise date T	7
$\tilde{P}_K(t, T)$	Price at t of American put	7
$P^{F_K}(t, T)$	Price at t of put on forward contract	7
$P_t[T_0, T_n]$	Swap pv01 at t for swap of period $[T_0, T_n]$	4
q^*, Q^*	Risk-neutral probability, probability distribution with respect to alternate numeraire	9
r	Interest rate	1
S_t	Price of stock or other asset at t	1
t	Current time	2
T	Forward time, maturity or expiration	2, 7
$V^A(t)$	Value at t of portfolio **A**	6
$V_K(t, T)$	Value at t of forward with delivery price K and maturity T	2
	Value at t of FRA with fixed rate K and fixing date T	3
$V_K^{SW}(t)$	Value at t of swap with fixed rate K	4
$V_K^{FXD}(t)$	Value at t of fixed leg of swap with fixed rate K	4
$V^{FL}(t)$	Value at t of floating leg of swap	4
W	Standard normal random variable, $W \sim N(0, 1)$	10
W_t	Brownian motion	16
$X \sim N(\mu, \psi^2)$	X is normally distributed with mean μ and variance ψ^2	9
$Y \sim \text{lognormal}(\mu, \sigma^2)$	$\log Y$ is normally distributed with mean μ and variance σ^2	10
$y_t[T_0, T_n]$	Forward swap rate at t for swap of period $[T_0, T_n]$	4
$\widetilde{y}_t[T_0, T_n]$	Constant maturity swap rate at t	14
$Z(t, T)$	Price at t of zero coupon bond with maturity T	1

Symbol	Definition	Chapter
α	Accrual factor in fractions of years	1
σ	Volatility, standard deviation of logarithm of asset	10
σ_K	Implied volatility for option of strike K	11
$\sigma(t,T)$	Volatility at t for libor forward with maturity T	15
$\phi(x)$	Standard normal probability density function	9
$\Phi(x)$	Standard normal cumulative distribution function	9
$\Phi\,(t,T)$	Futures price at t for futures contract with maturity T	5
ψ	Standard deviation of asset	9
$\Psi_K(t,T_0,T_n)$	Price at t of K-strike European payer swaption with exercise date T_0 on swap from T_0 to T_n	12

Abbreviations

		Chapter
ATMF	at-the-money-forward	7
BGM	Brace–Gatarek–Musiela	15
CMS	constant maturity swap	14
FRA	forward rate agreement	3
ZCB	zero coupon bond	1
3mL	three-month libor	3

Black–Scholes formula

$$C_K(t,T) = S_t\Phi\,(d_1) - Ke^{-r(T-t)}\Phi\,(d_2)$$

$$\text{where } d_1 = \frac{\log\left(\frac{S_t}{K}\right) + \left(r + \frac{1}{2}\sigma^2\right)(T-t)}{\sigma\sqrt{T-t}} \text{ and } d_2 = d_1 - \sigma\sqrt{T-t}.$$

REFERENCES

Black, F. (1976). The pricing of commodity contracts. *Journal of Financial Economics*, 3, 167–79.

Black, F. and Scholes, M. (1973). The pricing of options and corporate liabilities. *Journal of Political Economy*, 81 (3), 637–54.

Blyth, S.J. (2004). Practical relative value volatility trading. *RISK*, 17 (5), 91–6.

Blyth, S.J. (2011). The quant delusion. *RISK*, 23 (1), 121–3.

Blyth, S.J. (2012). The quant delusion: financial engineering in the post-Lehman Dodd-Frank landscape. *CFA Institute Conference Proceedings Quarterly*, 29 (1), 1–8.

Brace, A. Gatarek, D. and Musiela, M. (1997). The market model of interest rate dynamics. *Mathematical Finance*, 7(2), 127–154.

Corb, H. (2012). *Interest rate swaps and other derivatives.* Columbia University Press, NY.

Dupire, B. (1994). Pricing with a smile. *RISK*, 7 (1), 18–20.

Derman, E. and Kani, I. (1994). Riding on a smile. *RISK*, 7 (2), 32–9.

Harrison, J.M. and Kreps, D.M. (1979). Martingales and arbitrage in multiperiod securities markets. *Journal of Economic Theory*, 20, 381–408.

Hull, J.C. (2011). *Options, Futures and Other Derivative Securities* 8th ed. Prentice Hall, NY.

Hunt, P.J. and Kennedy, J.E. (2004). *Financial Derivatives in Theory and Practice* revised ed. Wiley, Chichester.

Joshi, M.S. (2008). *The Concepts and Practice of Mathematical Finance* 2nd ed. Cambridge University Press, Cambridge.

Miltersen, K. Sandmann, K. and Sondermann, D. (1997). Closed form solutions for term structure derivatives with lognormal interest rates. *Journal of Finance*, 52(1), 409–30.

Shreve, S.E. (2004) *Stochastic Calculus for Finance II: Continuous Time Models.* Springer, NY.

Taliaferro, R.D. and Blyth, S.J. (2011). Fixed income arbitrage in a financial crisis. *Harvard Business School*, Cases 211.049–52.

Williams, D. (1991). *Probability with Martingales.* Cambridge University Press, Cambridge.

Wilmott, P. (1998). *Derivatives: The Theory and Practice of Financial Engineering.* Wiley, Chichester.

INDEX